CONTENTS

1. INTRODUCTION 7

2. PRETREATMENT OF SAMPLES 11
   2.1 Dilution of samples 11
   2.2 Concentration of samples 11
   2.3 Handling procedures 14
   2.4 Permanent preparations and preservation of
       microorganisms 16

3. DIRECT COUNTS ON MEMBRANE FILTERS 19
   3.1 Summary of the method 19
   3.2 The membranes 19
       3.2.1 Cellulosic membranes 19
       3.2.2 Polycarbonate membranes 20
       3.2.3 Black membranes for epifluorescence microscopy . 20
       3.2.4 Electron microscopy 21
   3.3 Filtration 22
       3.3.1 The apparatus 22
       3.3.2 The procedure 22
   3.4 Microscopy and enumeration 24
       3.4.1 Examination of the unstained membrane . . . 24
       3.4.2 Examination of material stained with erythrosine . . 25
       3.4.3 Incident light fluorescence (epifluorescence) methods 25
   3.5 The organisms 29
   3.6 Analysis of results 30

4. THE USE OF COUNTING CHAMBERS 33
   4.1 Chambers for smaller microorganisms (0.5 μm to 10 μm) . 33
   4.2 Chambers for larger microorganisms ($> 10\,\mu$m) 36
       4.2.1 Sedimentation chambers 36
       4.2.2 The Sedgewick-Rafter cell 39
   4.3 Simple chambers which may be made in the laboratory . . 39
       4.3.1 Moat chamber for larger microorganisms . . 39
       4.3.2 Lund chamber 40
       4.3.3 Marked slide 40
   4.4 Counting without chambers 42
   4.5 Microscopy 43
   4.6 The organisms 43

BIBLIOGRAPHY

MANAGEMENT ACCOUNTING, PRODUCTIVITY REPORT. London. *The Anglo-American Council on Productivity*, 1950. 71 pages.

MANAGEMENT ACCOUNTING, A CONCISE APPRAISAL. London. *The Association of Certified and Corporate Accountants*. 19 pages.

MANAGEMENT ACCOUNTING FOR THE SMALL BUSINESS. London. *The Association of Certified and Corporate Accountants*. 1963. 35 pages.

DEVELOPMENTS IN MANAGEMENT ACCOUNTANCY. J. Batty, ed. London. *Heinemann/The Institute of Cost and Management Accountants*, 1968. 507 pages.

MANAGEMENT ACCOUNTANCY. J. Batty. London. *MacDonald and Evans*, 1970. Third edition. 875 pages.

SIMPLIFIED MANAGEMENT ACCOUNTING FOR SMALLER FIRMS. London. *The British Institute of Management*, 1952. 18 pages.

ACCOUNTS FOR MANAGEMENT, SOME ASPECTS OF FINANCIAL CONTROL IN SMALL AND MEDIUM SCALE BUSINESS. F. C. De Paula. London. *The British Institute of Management*, 1965. Revised edition. 83 pages.

MANAGEMENT ACCOUNTING IN PRACTICE. F. C. De Paula. London. *Pitman*, 1967. Third edition. 216 pages.

FINANCE AND ACCOUNTS FOR MANAGERS. Desmond Goch. London. *Pan Books*, 1971. Revised edition. 186 pages.

AN INTRODUCTION TO BUDGETARY CONTROL, STANDARD COSTING, MATERIAL CONTROL AND PRODUCTION CONTROL. *The Institute of Cost and Management Accountants*, 1950.

PLANNING IN THE SMALL FIRM. GUIDELINES FOR THE SMALLER BUSINESS NO. 2. London. *The British Institute of Management*, 1969.

BUSINESS FORECASTING. London. *The British Institute of Management*, 1949. 32 pages.

AN INTRODUCTION TO BUSINESS FORECASTING. London. *The Institute of Cost and Management Accountants*, 1960. 42 pages.

TERMINOLOGY OF COST ACCOUNTANCY. Third edition, London. *The Institute of Cost and Management Accountants. Gee & Co (Publishers) Ltd*, 1966. 29 pages.

Five weeks to

Products	
A	
B	
C	
D	
E	
F	
Chargeable tools	
	£

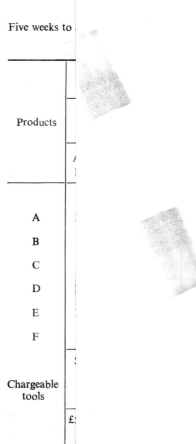

each product group because, in this case, it is preferred to show the effect of the sales mix for all product groups.

Because a MANAGEMENT ACCOUNTING system is tailored to meet the structure of an organization and the people working in it, it is not of scientific or technical design but is more in the nature of an art form. Like society and industry, it reflects a biological pattern.

As imaginative research and investigation is employed to find the weaknesses, strengths and potentials of the structure and people, further developments of the management accounting system must take place.

This does not necessarily mean added information or greater complexity – such developments should be viewed with suspicion – the aim must be towards obtaining and presenting information which will be more effective in managing and stimulating the organization as it develops.

It is in this creative approach to management accounting that the management accountant will make his greatest contribution.

sales prices and at works standard costs, a fairly substantial undertaking.

It is essential to see how profits are being affected by variations from budget of the prices, volume and mix of actual sales and this information is shown in the last four columns of the revised statement of sales and margins (Exhibit No. 27).

It will be noticed that the statement, as illustrated, is a working paper which contains certain columns necessary to calculate the variances shown in the last four columns. On the statement circulated to the directors and the sales department, the columns used to calculate the variances are omitted and only those ticked at the foot are shown.

The method of preparing the statement is first to enter in column 1 the actual sales at actual prices as shown by the sales invoices. The actual sales at standard prices are then entered in column 2 and the corresponding works standard costs of those sales are entered in column 4; this information is obtained by recording the standard sales price and the works standard cost of each item appearing on the sales invoices. The extensions and summarizing are done on a calculating machine.

The budgeted sales at standard prices (column 3) and the works standard cost of budgeted sales (column 5) are then completed. As the statement covers a five-week period, the amounts appearing in these columns are obtained by taking 5/52 weeks of the corresponding items shown in the sales budget (Exhibit 17). This statement shows for the period which it covers that the total actual sales produced a gain in gross margin of £258 (column 12) and that this total variance is made up as follows:

		£
(a) Actual prices being more favourable than budget	+	62 (column 13)
(b) The actual volume of sales being less than budget	−	1,003 (column 15)
(c) The mix of actual sales being more favourable than budget	+	1,199 (column 15)
		£258

The statement also shows for each of the main product groups the total variance (column 12), the price variance (column 13), and the volume and sales mix variance (column 14). The latter variance is not analysed between volume and sales mix for

The extra allowances and efficiency variances are calculated as follows:

(a) *Extra allowances variance*
This is the cost of the allowances granted at the standard rate of pay, thus:
100 castings at 2 minutes each = 3 hours 20 minutes at 30p per hour .. £1·00

(b) *Efficiency variance*
This is the difference between the standard time plus the extra allowances and the actual time taken at the standard rate of pay, thus:

Standard time = 100 castings at 12 minutes each	20 hrs
Extra allowance = 100 castings at 2 minutes each	3 hrs 20 mins
	23 hrs 20 mins
Actual time taken	18 hrs
Saving of time	5 hrs 20 mins
Efficiency variance (favourable) = 5 hrs 20 mins at 30p per hour	£1·60

The extra allowances are usually claimed by operatives and the claims forwarded to the planning department by their foremen. The planning department assesses the extra time to be allowed for the additional work not envisaged at the time of setting the standard time for the job.

To keep the examples simple, those shown above were based on a specific job. The total variances for the various production departments are arrived at by aggregating the same data for all the jobs which constitute the production for the department for one period.

Sales variances

The statement of sales and margins (Exhibit No. 4) described in Chapter III does not analyse the total sales variance into sales price, sales volume and sales mix variances; to do so would have involved analysing the sales invoices both at standard

Wages – efficiency variance

At present the total wage variances in the production departments are analysed between rates of pay variance and efficiency variance. A further analysis of the efficiency variance can show:

(a) the loss due to extra allowances being granted for work on account of special difficulties being incurred by the operator, e.g., faulty materials or errors in machine setting;

(b) the gain or loss arising from work taking less or more time than provided for in the product standard cost plus the extra allowances, i.e., the normal efficiency variance.

This analysis is arranged mainly to bring out the amount of extra allowances granted; these allowances would otherwise be absorbed in losses arising from working below standard efficiency or masked by gains arising from increased efficiency in working.

The nature and method of calculating the extra allowances and efficiency variances are illustrated below.

Assume:

(i) Standard time for machining a casting – each .. .. 12 mins
(ii) Extra allowance granted by the planning department to cover extra work due to unusually rough castings – each .. 2 mins
(iii) Number of castings 100
(iv) Standard rate of pay – per hour .. 30p
(v) Time taken on 100 castings .. .. 18 hrs

The total variance (formerly efficiency variance) is determined as follows:

Standard time for machining 100 castings
100×12 minutes each .. .. .. 20 hrs
Actual time taken .. .. 18 hrs

Total time saved .. .. 2 hrs

The total variance is therefore the total time saved $\times$ the standard rate of pay $= 2$ hours $\times 30p$ 60p

REFINEMENTS AND DEVELOPMENTS OF THE SYSTEM

Assume:

(i) Fixed expenses per budget	£1,000
(ii) Standard hours allowed for budgeted production	2,000
(iii) Standard fixed overhead rate per standard hour (£1,000÷2,000 hours)	50p
(iv) Standard hours allowed for actual production	1,500
(v) Actual hours taken for actual production	1,200

The total volume variance is arrived at as follows:

	£
Fixed expenses recovered in standard costs of production = 1,500 hours at 50p per hour	750
Budgeted fixed expenses	1,000
Total volume variance – unfavourable	£250

The efficiency and capacity variances are calculated as follows:

(a) *Efficiency variance*:

	£
Fixed expenses recovered in the standard costs of actual production = 1,500 hours at 50p per hour	750
Fixed expenses – calculated recovery based on actual production hours = 1,200 hours at 50p per hour	600
Efficiency variance – favourable	£150

(b) *Capacity variance:*

	£
Fixed expenses – calculated recovery based on actual production hours = 1,200 hours at 50p per hour	600
Fixed expenses – calculated recovery based on budgeted standard hours = 2,000 hours at 50p per hour	1,000
Capacity variance – unfavourable	£400

This example brings out the fact that the under-utilization of the production capacity of the department is more serious than is disclosed by the total volume variance. By analysing the volume variance in this way, it becomes clear that the under-recovery of fixed overheads due to the department being under-employed is being offset to some extent by more efficient working.

allocate responsibility for the variances to the correct centres; and (c) to assist the persons in charge of those centres to exercise more effective control of their costs.

Secondly, the variances themselves may be subdivided. For example, the volume variance relating to overhead expenses may be analysed between (i) efficiency variance; and (ii) capacity variance. The material variance may be analysed between (i) material usage variance; and (ii) change of specification variance. This analysis of the original variances provides additional information as to the causes of divergencies from standard, and narrows the field for any inquiries which may be necessary.

The plans for developing the original variances of this are described below; as will be seen they are not extensive.

Expenses – volume variance

When labour costs based on standard operating times are introduced in place of the present piece-work rates, it is possible to analyse the volume variance in each production department to indicate how much of the variance is due to the efficiency of working in the department, and how much is due to the extent to which its production capacity is being utilized.

This means the volume variance could be analysed into:

(a) *Efficiency variance* which is the gain or loss arising from the actual time taken being less or more than the standard time allowed for the actual production achieved.

(b) *Capacity variance* which is the gain or loss arising from the actual time for the actual production achieved being less or more than the standard time of budgeted production as determined by the production budget.

Where, as in the case of this particular company, overhead expenses are considered to be either fixed or variable, the volume variance arises (as explained in Chapter III) because the fixed overhead expenses recovered in the cost of actual production are greater or less than the recovery based on the budgeted production. The following example illustrates how the volume variance is analysed to show the extent to which it is due to the efficiency of working in the department (efficiency variance) and the employment of the productive capacity of the department (capacity variance).

REFINEMENTS AND DEVELOPMENTS OF THE SYSTEM 95

rate will then be obtained by multiplying the standard time by the standard hourly rate.

It will then be possible for the overhead rates of the production departments to be expressed as a cost per unit of standard time, for example, 50p per standard hour or 1p per standard minute instead of a percentage of the standard labour cost. It follows that overheads will then be charged in the cost of production at a rate per standard hour or minute.

The method of charging overheads on the basis of standard hours rather than on standard labour cost is more equitable, as it removes the anomaly in the labour cost method that the more skilled and highly paid operations attract more overheads than the less skilled and less well-paid operations.

The introduction of standard times will enable the efficiency of labour to be more readily assessed for operations, individuals and departments by a direct comparison of the time allowed and the time taken.

Another advantage of standard times is that production capacity can be expressed in standard hours. It will then be possible, by introducing a capacity variance, to show the extent to which the capacity of the departments is being utilized; this variance is described in the next section.

Development of variances

The variances shown on the profit and loss account and operating statements described in Chapter III are those which, at the outset of the scheme, were considered to be the minimum necessary to enable (*a*) a reasonable measure of control to be exercised over the costs and operations of the business, and (*b*) the standard cost of stocks and work in progress to be adjusted to 'actual' for balance sheet purposes.

Where, however, a scheme is started off with only those variances, it is usually desirable within a short time to develop the original scheme of variances in a number of ways.

First, the departments as originally established can be broken down into budget or cost centres consisting of either (i) a specific area of responsibility, e.g., a manufacturing section under a charge-hand; or (ii) a specific type of operation or function, e.g., a group of machines of one kind or the export section of the sales department. It is then possible (*a*) to fix more precisely the point at which the variances arise; (*b*) to

The scrap variances in the departmental operating statement may be proved arithmetically as follows:

Operation No. 1	Casting 'A'		Casting 'B'	
Actual input	100 units		200 units	
Standard scrap allowance	10 per cent	10 ,,	16⅔ per cent	33·33 ,,
Actual scrap	(100−88)	12 ,,	(200−170)	30 ,,
Excess or saving on scrap	Excess	2 ,,	Saving	3·33 ,,
Standard cost of materials including scrap allowance, per unit		40p		60p
Scrap variance		2 × 40p		3·33 × 60p
		=Excess 80p		=Saving 200p

Operation No. 2				
Actual input	88 units		170 units	
Standard scrap allowance	5 per cent	4·4 ,,	10 per cent	17 ,,
Actual scrap	(88−84)	4·0 ,,	(170−150)	20 ,,
Excess or saving on scrap	Saving	0·4 ,,	Excess	3 ,,
Standard cost of materials including scrap allowance, per unit		60p		100p
Scrap variance		0·4 × 60p		3 × 100p
		=Saving 24p		=Excess 300p

Accounting for scrap in this way does involve more work in preparing product standard costs than the original method. The introduction of scrap allowances and variances should provide a degree of control over the cost of scrap which does not otherwise exist.

Standard times

Standard labour rates are established for each operation as explained in Chapter VII on the basis of the actual piece-work price for the job to which is added a percentage to cover the hourly bonus rate. A system of standard times for labour costing and wage payment could be used as an alternative.

To do this department by department, work will be time-studied and a standard time established for each operation. Standard hourly rates will also be established for the grade of labour to be used on each operation and the standard labour

These costings are then used for evaluating actual cost and the standard cost of production of the department. The following detailed extract from a departmental operating statement shows how these figures are used so far as materials and scrap are concerned.

Detailed extract from a departmental operating statement
Department: MACHINE SHOP

Production		Casting 'A'	Casting 'B'
Input		100	200
Output – Operation No. 1 ..		88	170
Output – Operation No. 2 ..		84	150

Materials	Actual cost	Standard cost	Scrap variance
Operation No. 1	p	p	p
Casting 'A' ..	(100 × 36p) 3,600	(88 × 40p) 3,520	80*
Casting 'B' ..	(200 × 50p) 10,000	(170 × 60p) 10,200	200
	13,600	13,720	120
Operation No. 2			
Casting 'A' ..	(88 × 57p) 5,016	(84 × 00p) 5,040	24
Casting 'B' ..	(170 × 90p) 15,300	(150 × 100p) 15,000	300*
	20,316	20,040	276*
	33,916p	33,760p	156p*

(*unfavourable*)

In Operation No. 1 the 'actual' cost represents the rough castings issued from the stores and would be debited to the departmental operating account and credited to the stock of raw materials account. The standard cost of Operation No. 1 represents the standard materials cost of the good output (per job tickets) and, together with labour and overheads at standard cost, is credited to the departmental operating account and debited to the work in progress account. On the assumption that Operation No. 2 has taken place within the accounting period, the 'actual' cost of the materials used for this operation corresponds to the total standard cost of Operation No. 1 and would be credited to the work in progress account and debited to the departmental operating account. The standard cost of Operation No. 2 represents the standard material cost of the good output (per job tickets) and, together with labour and overheads at standard cost, is credited to the departmental operating account and debited either to a finished stock account, or to another work in progress account if further operations are necessary.

SUMMARIZED PRODUCT STANDARD COSTS

Dept.: Machine Shop

	Casting 'A'		Casting 'B'	
Operation No. 1		p		p
1. Raw materials – 100 castings	at 36p each	3,600	at 50p each	5,000
2. Percentage scrap on input	10 per cent		$16\frac{2}{3}$ per cent	
3. Uplift on materials cost: $\left(\dfrac{\text{Item 2}}{100-\text{Item 2}}\right)$ ..	$\dfrac{10}{100-10}=\cdot1111$		$\dfrac{16\frac{2}{3}}{100-16\frac{2}{3}}=\cdot2$	
4. Uplift applied to materials cost: (Item 3 × Item 1)	$\cdot1111\times3{,}600\text{p}$	400	$0\cdot2\times5{,}000\text{p}=$	1,000
5. Raw materials – 100 partly finished castings	(40p each)	4,000	(60p each)	6,000
6. Direct labour on 100 good castings ..	4·25p per Cstg	425	7·5p per Cstg	750
7. Overheads on direct labour – 300 per cent		1,275		2,250
	(57p each)	5,700p	(90p each)	9,000p
Operation No. 2		p		p
8. Materials from Operation No. 1 – 100 partly finished castings		5,700		9,000
9. Percentage scrap on input	5 per cent		10 per cent	
10. Uplift on materials cost: $\left(\dfrac{\text{Item 9}}{100-\text{Item 9}}\right)$ ..	$\dfrac{5}{100-5}=\cdot0526$		$\dfrac{10}{100-10}=\cdot1111$	
11. Uplift applied to materials cost: (Item 10 × Item 8)	$\cdot0526\times5{,}700\text{p}$	300	$\cdot1111\times9{,}000\text{p}=$	1,000
12. Materials – 100 finished castings ..	(60p each)	6,000	(100p each)	10,000
13. Direct labour on 100 good castings ..	10p per Cstg	1,000	12p per Cstg	1,200
14. Overheads on direct labour – 300 per cent		3,000		3,600
15. Total standard cost of 100 units ..	(100p each)	10,000p	(148p each)	14,800p

will cease to be charged as a departmental overhead; instead it will be included as a percentage addition to the standard cost of the materials in each operation. The following example illustrates the proposed new method of costing for scrap; it consists of summarized product standard costs for two castings which pass through two machining operations:

(if a piece-work job) and the operation to be performed; the time of starting and finishing the job is recorded by a time clock and, as each operation is completed, the ticket is signed by an inspector. The tickets are then passed to the wages department where they are used to calculate gross earnings. Job tickets can be 'preprinted' and show, in addition to the information mentioned above, the standard labour cost of each operation. After gross earnings have been calculated by the wages department the job tickets can be machine listed by the cost office to obtain the standard labour cost of completed operations for each production department. The departmental overhead rates can then be applied to the standard cost of labour in each department to obtain the standard labour and overhead cost of production.

Materials requisitions can also be preprinted by the planning department. In addition to showing the standard cost of raw materials required for each operation, the requisitions can show the standard cost of bought-out parts and sub-assemblies. Formerly, bought-out parts and sub-assemblies were charged direct to work in progress and not to the production departments. Preprinted material requisitions can, however, save work in the cost office and it is therefore proposed to charge each department with the standard cost, not only of raw materials as previously, but also of bought-out parts and sub-assemblies, including partly processed materials, and parts and sub-assemblies from other departments. The corresponding credit for production will then be for the full cumulative standard cost of production and not merely the standard cost of the production by the department.

The cost office can machine list the requisitions to obtain the total standard cost of materials issued to each department; it can then summarize the standard cost of labour, overheads and materials to obtain the standard cost of production for each department.

Scrap

As explained in Chapter VII, the standard cost of scrapped work is charged as an overhead expense to the department responsible for the scrap. The development here is that the planning department can establish standard scrap allowances at each operation. When this has been done, the cost of scrap

overheads and profit, the sales manager is in a position to see the effect of varying his prices and quantities. He will have to stay, of course, within the bounds of practical possibility and he will have to assume, for the time being, that the direct cost of the products and the fixed overheads of the business will remain the same. He may therefore revise his first forecast as follows:

	Product A	Product B	Product C
Quantity	16,000	8,000	7,000
Price, each	£1·25	£2·25	£3·50
Turnover	£20,000	£18,000	£24,500
Direct cost, each	47·5p	£1·25	£1·50
Contribution, each	77·5p	£1	£2
Total contribution	£12,400	£8,000	£14,000

The total turnover of this revised forecast is £62,500 and the total contribution to fixed overheads and profit £34,400. If the fixed overheads remained at £25,000, the net profit for the year would now be £9,400.

The revised forecast therefore shows an increased profit of £3,900 and, although rather in the nature of a textbook exercise, it does show that by knowing the contribution made by each product the sales manager is in a better position to plan sales with a view to achieving the maximum net profit. However, it must be remembered that many companies have a more or less established pattern and level of production, which cannot be changed significantly without affecting production facilities; therefore there are limits to the extent to which the mix, quantities and prices of sales can be varied.

Standard cost of production

As explained in Chapter VII the finished production of each department as reported by the progress office is evaluated at the standard cost of raw materials, direct labour and works overhead; this information is taken from the products standard cost cards.

This evaluation of finished work only is not always satisfactory and it is often desirable that credit be given to the departments for the uncompleted work. In such a situation it is necessary to value production by reference to operations as completed and recorded on job tickets.

Job tickets are issued for every job to be undertaken, giving details of the department, quantity required, piece-work rate

It is becoming increasingly important, when considering sales prices and the relative profitability of products, to have this information available to be able to see the contribution made by each product to fixed overheads and profit. It is desirable, therefore, to show separately on the product standard cost cards the rates and amounts of fixed and variable overheads. For example, the product standard cost illustrated on Exhibit No. 16 might be summarized as follows:

	p	p
Standard selling price		240·00
Direct cost:		
Bought-out parts	27·00	
Raw materials	39·00	
Direct labour	24·50	
Works variable overheads	25·25	
		115·75
Contribution to fixed overheads and profit		124·25p
Analysis of contribution:		
Works fixed overheads		38·40
Selling and administration expenses		25·24
Profit		60·61
		124·25p

As will be seen, the total of the materials, labour and works variable overheads is termed direct cost, and the difference between the direct cost and the standard selling price is considered to be the contribution towards fixed overheads and profit. This information will be of value to the sales manager when deciding such matters as the prices and quantities of sales and the direction of sales effort.

He may, for example, provisionally forecast that during the year he can sell:

	Product A	Product B	Product C
Quantity	20,000	10,000	5,000
Price each	£1	£2	£4
Turnover	£20,000	£20,000	£20,000
Direct cost, each	£47·5p	£1·25	£1·50
Contribution, each	£52·5p	£75·0p	£2·50
Total contribution	£10,500	£7,500	£12,500

The total turnover from this forecast is therefore £60,000 and the total contribution to fixed overheads and profit £30,500. If the fixed overheads were £25,000, the net profit for the year would be £5,500.

In the light of the contribution made by each product to fixed

accountant becomes so valuable and necessary. The accountant is the member of the executive who fully understands the principles involved and is therefore able to teach and guide the other members of the organization as to the work required.

The accountant may also use the advice of higher management, economists, buyers and personnel officers to estimate future changes in volume and expenses which may occur during the period covered by the budget but which may not be known by a departmental head.

Flexible budgets and allowed expenses
A further development in the budgeting for expenses will be to introduce a system of flexible budgeting.

As already explained, the present management accounting scheme assumes that overhead expenses are either fixed or variable. It follows that the allowed expenditure for a period is the sum of the budgeted fixed expenses, and the budgeted variable expenses; the variable expenses are adjusted proportionately by the percentage of production achieved.

Under a system of flexible budgeting, departmental expenses will be budgeted for several levels of production which might be, for example, 50, 75, 125 and 150 per cent of the production used for the official budget. The estimated amount of expense at the various production levels would then be plotted on a graph. The actual production for a period may be somewhere between these selected levels. If it was, say, 110 per cent, the amount of expense allowed to the department at this level would be read off the corresponding point on the graph.

The department overhead rates would continue to be calculated by reference to the official budgeted level of production. The analysis of the total expense variance between volume and spending would, however, be more realistic as the allowed expense, upon which this variance analysis depends, would be more accurate.

Overhead rates
The overhead rates for the various departments consist of the budgeted overhead expenditure expressed as a percentage of the budgeted direct labour. These rates do not distinguish between fixed and variable overheads and consequently neither do the product standard costs.

CHAPTER IX

REFINEMENTS AND DEVELOPMENTS
OF THE SYSTEM

The system of management accounts described in Chapters VII and VIII is sufficient to provide valuable information at determined intervals to all levels of management.

There are steps, however, which can improve the effectiveness of the system.

Preparation of expenses forecasts and budgets
In previous chapters it has been said that the accountant prepared budgets in conjunction with various members of management. This statement is not a very positive one and if the situation is not enlarged upon, then the system of control is likely to lose dynamism.

A forecast or a budget should always be made by the person responsible for the department or organization the activities of which are measured by the budget.

This is not to say that the budgetor, to coin a phrase, may not be dependent on the accountant for factual information, measurement and extrapolation.

Also, the budgetor may find his assessments and targets subjected to higher authority and the board. In the end, however, unless the budget receives the approval and support of the person who has to work to it, then co-operation and drive will be missing from the organization.

A total budget is not unlike a jigsaw puzzle where every part must be fitted in the right place to obtain the total picture required.

Every section of the business, every department, every cost centre must prepare its activity figures and its expenses, although the whole must be related within an agreed framework and dimension. For example, sales turnover cannot be agreed except in relation to production or purchase potential; production cannot be agreed except in relation to sales potential, labour available and material supplies.

It is in the fitting of the jigsaw together that the work of the

THE INTEGRATED ACCOUNTING SYSTEM

section of the general ledger to the profit and loss account, except the self-balancing groups of accounts which are dealt with in posting No. VII.

After this posting the balance on the profit and loss account in the ledger will agree with the profit or loss shown by the 'financial' accounts for the year.

Posting No. VII – Closing off self-balancing accounts in the profit and loss section of the general ledger

This posting closes off all the accounts in the self-balancing groups of accounts in the profit and loss section of the general ledger.

The remaining balances on the general ledger are now only balance sheet accounts.

Exhibit No. 26

ILLUSTRATION OF STOCK AND WORK IN PROGRESS ACCOUNTS AT THE BEGINNING AND END OF THE FINANCIAL YEAR

STOCK OF RAW MATERIALS (07.01)

Beginning of Period 1	Balance (per balance sheet) Stock and work in progress reserve	I	1,000 50	Beginning of Period 1	Balance, c/d ..		1,050
			£1,050				£1,050
Beginning of Period 1 Periods 1–12	Balance (at standard prices) b/d .. Purchases of raw materials at standard prices (an entry each period)		1,050 11,000	Periods 1–12 End of Period 12	Issues of raw materials at standard prices (an entry each period) .. Balance, c/d ..		10,650 1,400
			£12,050				£12,050
End of Period 12	Balance, b/d ..		1,400	End of Period 12 ,,	Stock and work in progress adjustment .. Balance, c/d ..	II	120 1,280
			£1,400				£1,400
End of Period 12	Balance, b/d .. Stock and work in progress reserve	III	1,280 190	End of Period 12	Balance, c/d ..		1,470
			£1,470				£1,470
Beginning of Period 1	Balance (per balance sheet) b/d		1,470				

WORK IN PROGRESS (07.02)

Beginning of Period 1	Balance (per balance sheet) Stock and work in progress reserve	I	3,300 230	Beginning of Period 1	Balance, c/d ..		3,530
			£3,530				£3,530
Beginning of Period 1 Periods 1–12	Balance (at standard prices), b/d Purchase of bought-out parts at standard prices (an entry each period) .. Production transferred from production depts at standard cost of direct labour, raw materials and expenses (an entry each period) ..		3,530 4,000 23,500	Periods 1–12 End of Period 12	Standard cost of sales (an entry each period) .. Balance, c/d ..		28,430 2,600
			£31,030				£31,030
End of Period 12	Balance, b/d ..		2,600	End of Period 12	Stock and work in progress adjustment .. Balance, c/d ..	II	2,? 2,?
			£2,600				£2,?
End of Period 12	Balance, b/d .. Stock and work in progress reserve	III	2,400 80	End of Period 12	Balance, c/d ..		2,?
			£2,480				
Beginning of Period 1	Balance (per balance sheet), b/d		2,480				

STOCK AND WORK IN PROGRESS RESERVE (07.04)

End of Period 12	Consumable stores and other stocks Profit and loss account	IV V	100 450	Beginning of Period 1 End of Period 12	Stock of raw materials and work in progress .. Stock of raw materials and work in progress ..	I III	280 270
			£550				£550

STOCK AND WORK IN PROGRESS ADJUSTMENT (07.05)

End of Period 12	Stock of raw materials and work in progress ..	II	500	End of Period 12	Profit and loss account ..	V	500
			£500				£500

(The figures in the folio columns refer to the corresponding posting numbers on Exhibit No. 25.)

from, the standard cost the material price and operating variances which have arisen during the period in which stocks and work in progress have been accumulated. For example, if the stock of raw materials had been accumulated over a period of, say, four months, then the price variances which had arisen during that period would be added to, or deducted from, the standard cost of the stocks. If the price variances were unfavourable, i.e., higher prices than standard had been paid, they would be added to the standard cost of stocks; conversely, favourable variances would be deducted.

Similarly, unfavourable operating variances during the 'build-up' period would be added to the standard cost of work in progress and favourable variances deducted.

The 'actual' cost of stocks and work in progress is then written down for balance sheet purposes wherever such factors as obsolescence or a fall in market prices below cost prices make this necessary.

The total difference between the standard cost valuation and the balance sheet valuation is then transferred by this posting to the stock and work in progress reserve account.

Posting No. IV – Transfer of differences arising on valuation of closing consumable stores and other stocks for balance sheet purposes

As with production stocks, the actual values of consumable stores and other stocks are written down, wherever necessary, to balance sheet values and the differences transferred to the stock and work in progress reserve account.

Posting No. V – Transfer of stock and work in progress adjustment and reserve accounts

This posting closes off the stock and work in progress adjustment and reserve accounts to the profit and loss account; these transfers are shown as special items on the 'management' profit and loss account for the last period of the year, and the cumulative profit or loss shown by that account will then agree with that shown by the 'financial' accounts for the year.

Posting No. VI – Transfer of balances in the profit and loss section of the general ledger

This posting closes off all the balances in the profit and loss

Exhibit No. 25 – continued

		Account No.	£	£
End of Period 12	V *Transfer of stock and work in progress adjustment and reserve accounts* Dr. Profit and loss	23.01	50	
	Stock and work in progress reserve	07.04	450	
	Cr. Stock and work in progress adjustment	07.05		500
End of Period 12	VI *Transfer of balances in the profit and loss section of the general ledger* Dr. Sales	50.01 to 50.03	√	
	Miscellaneous income	66.01	√	
	Cr. Works cost of sales	52.01 to 52.02		√
	Standard cost of selling and administration expenses	62.01		√
	Miscellaneous expenditure	68.01		√
	Appropriations	70.01		√
	Dr. or Cr. Variances	64.01 to 64.06	√	√
	Cr. Profit and loss	23.01		√
End of Period 12	VII *Closing-off self-balancing accounts in the profit and loss section of the general ledger* Dr. Purchases of direct materials transfer	54.99	√	
	Materials used from stock and scrap made transfer	56.99	√	
	Wages and expenses transfer	58.99	√	
	Cr. Purchases of direct materials	54.01 to 54.02		√
	Materials used from stock and scrap made	56.01 to 56.03		√
	Wages and expenses	58.01 to 58.72		√

Exhibit No. 25

OPENING AND TERMINAL POSTINGS FOR FINANCIAL YEAR

		Account No.	£	£
	I *Transfer of difference arising on the valuation of opening production stocks at standard prices*			
Beginning of Period 1	Dr. Stock of raw materials	07.01	50	
	Work in progress	07.02	230	
	Cr. Stock and work in progress reserve	07.04		280
	II *Transfer of differences on stock and work in progress accounts following the year-end physical stock-taking*			
End of Period 12	Dr. Stock and work in progress adjustment	07.05	500	
	Cr. Stock of raw materials	07.01		120
	Work in progress	07.02		200
	Chargeable tools and dies work in progress	07.06		10
	Packing materials stock	07.07		40
	Consumable stores stock	07.08		30
	Consumable tools stock	07.09		50
	Printing and stationery stock	07.10		30
	Replacements to plant, machinery and equipment stock	07.11		20
	III *Transfer of difference arising on valuation of closing production stocks for balance sheet purposes*			
End of Period 12	Dr. Stock of raw materials	07.01	190	
	Work in progress	07.02	80	
	Cr. Stock and work in progress reserve	07.04		270
	IV *Transfer of differences arising on valuation of closing consumable stores and other stocks for balance sheet purposes*			
End of Period 12	Dr. Stock and work in progress reserve	07.04	100	
	Cr. Packing materials stock	07.07		30
	Consumable stores stock	07.08		20
	Consumable tools stock	07.09		10
	Printing and stationery stock	07.10		20
	Replacements to plant, machinery and equipment stock	07.11		20

Postings at the beginning and end of the financial year
There are a number of postings made at the beginning and end of the financial year. These postings relate mainly to the stock and work in progress accounts and are made to reconcile the traditional 'financial' profit and loss account with the 'monthly' profit and loss accounts in standard cost form. These special postings are shown in the form of journal entries on Exhibit No. 25, and the stock and work in progress accounts concerned are shown on Exhibit No. 26. Figures have been included on these exhibits to help in tracing the postings from the cost journal to the ledger accounts.

Posting No. I –   Transfer of the difference arising on the valuation of opening production stocks at standard prices

At the beginning of the year the opening production stocks (including work in progress) are valued at the standard prices to be used for the ensuing year. The difference between the valuation at the standard prices and the balance sheet valuation is transferred by this entry to the stock and work in progress reserve account.

Posting No. II –   Transfer of differences on stock and work in progress accounts following the year-end physical stock-taking

When physical stock-taking is done at the end of the year, the production stocks are valued at standard prices; consumable stores and other stocks are valued at actual prices as standard prices have not been established for these materials. The total values of the stocks should agree with the total values shown by the various stock and work in progress accounts in the general ledger. In practice there are differences and these, after investigation, are transferred by this entry to the stock and work in progress adjustment account.

Posting No. III –   Transfer of difference arising on valuation of closing production stocks for balance sheet purposes

The standard cost of stocks and work in progress is then adjusted to arrive at an 'actual' cost for balance sheet purposes. The 'actual' cost may be determined by adding to, or deducting

Continued on page 84

Exhibit No. 24 – continued

that labour. This volume variance may not, however, be a realistic reflection of the changes in the volume of work done by the tool-room as it may be that by more efficient working a smaller staff is doing as much work as the larger staff budgeted for. This presentation of the variances in the tool-room was, however, considered to be the best at this stage. At a later stage it may be possible to establish standards of work for the tool-room so enabling a more realistic volume variance to be shown.

Selling and administration variances

The summary of service, selling and administration departments expenses (Exhibit No. 8) shows an unfavourable variance of £1,910 on selling and administration for period 6.

The allowed expense of £6,706 is equal to the budgeted figure for the period. The budgeted expenses for selling and administration are expressed as a percentage of works standard cost of budgeted sales to arrive at a selling and administration overhead rate of 16·38 per cent.

The standard recovery of £6,310 is equal to the selling and administration overhead rate of 16·38 per cent applied to the works standard cost of actual sales of £38,524 (Exhibit No. 4).

The actual expenses are £8,200.

The total variance, £1,910 unfavourable, is the difference between standard recovery of £6,310 and actual expenses of £8,220.

The analysis of total variance is therefore:

(a) *The volume variance (£396 unfavourable)*

The volume variance is the difference between	
standard recovery of	£6,310
and allowed or budgeted expenses of	6,706
	=£396 unfavourable

and represents the selling and administration overhead rate of 16·38 per cent applied to the fall of £2,418 from the works standard cost of budgeted sales of £40,942 to the works standard cost of actual sales of £38,524.

(b) *The spending variance (£1,514 unfavourable)*

The spending variance is the difference between	
allowed or budgeted expenses of	£6,706
and actual expenses of	8,220
	=£1,514 unfavourable

Exhibit No. 24 – continued

	1	2 Standard recovery (equal to actual tool-room labour plus recovery rate of 86·68%)	3	4=2−3	5=2−1	6=1−3
	Allowed expenses equal to budget £	£	Actual expenses £	Total variance £	Volume variance £	Spending variance £
Tool-room labour	3,769	2,764	2,764	—	1,005*	1,005
Tool-room expenses	3,267 (86·68% of labour)	2,396	3,408	1,012*	871*	141*
	£7,036	£5,160	£6,172	£1,012*	£1,876*	£864

(*unfavourable**)

The total variance, of £1,012, for this department arises only on the tool-room expenses as the whole of the labour is charged to those departments for which it is incurred; the total variance is the difference between standard recovery of £5,160 and actual expenses of £6,172.

The analysis of total variance is therefore:

(a) *Volume variance (£1,876 unfavourable)*
   The volume variance is equal to the standard recovery of £5,160, column 2, less the allowed or budgeted expenses of £7,036, column 1.

The amount of £1,876 represents the reduction in tool-room labour of £1,005 from the budgeted labour of £3,769 to the labour charged in other departments of £2,764 together with the overhead rate of 86·68 per cent on that reduction in labour amounting to £871.

(b) *The spending variance (£864 favourable)*
   The spending variance is equal to the allowed or budgeted expenses of £7,036 less the actual expenses of £6,172.

This represents a saving in the tool-room labour of £1,005 being the difference between the budgeted labour of £3,769 and the actual labour of £2,764 less the excess of the actual tool-room expenses of £3,408 over the budgeted expense of £3,267 amounting to £141.

Note

This method of treatment results in the tool-room labour showing an unfavourable volume variance offset by a favourable spending variance of the same amount. This is due to the fact that the tool-room wages actually paid are charged out to the departments using

Exhibit No. 24 – continued

Summary of Expenses and Variances

	1	2	3	4	5=3−4	6=3−2	7=2−4
					Total	Volume	Spending
	Budget	Allowed	Standard	Actual	variance	variance	variance
	£	£	£	£	£	£	£
Fixed expenses	788	788 (100% of budget)	630 (164% of £384)	863	233*	158*	75*
Variable expenses	1,132	906 (80% of budget)	906 (236% of £384)	905	1	—	1
Total	£1,920	£1,694	£1,536	£1,768	£232*	£158*	£74*
	(400% of £480)		(400% of £384)		(unfavourable*)		

The total variance of £232, unfavourable, in column 5 is equal to the standard expenses of £1,536 in column 3 less the actual expenses of £1,768 in column 4.

The analysis of total variance is therefore:

(a) *Volume variance (£158 unfavourable)*
   The volume variance of £158, unfavourable, in column 6 is equal to the standard expenses of £1,536, in column 3, less the allowed expenses of £1,694 in column 2. From the summary it can be seen that this is due to the under-recovery of the fixed expenses in actual production, i.e., actual production is 80 per cent of budget, therefore under-recovery is 20 per cent of fixed expenses, £788=£158.

(b) *Spending variance (£74 unfavourable)*
   The spending variance of £74, unfavourable, in column 7, is equal to the allowed expenses of £1,694, in column 2, less the actual expenses of £1,768, in column 4. This arises on both fixed and variable expenses and is equal to an excess of £75 on fixed expenses (allowed £788 less actual £863) less a saving of £1 on variable expenses (allowed £906, less actual, £905).

Service departments expenses variances

The summary of service, selling and administration departments expenses (Exhibit No. 8) shows an unfavourable variance on the tool-room of £1,012 for period 6.

The figures for allowed expenses, standard recovery and actual expenses may be analysed as follows:

Exhibit No. 24 – continued

Standard materials	*Actual materials*
Weight of zinc per materials specification plus allowance for melt loss in casting=17·385 lb. per item.	Quantity of zinc consumed (opening stock in department, plus issues from stores, less closing stock)= 70,118 lb.
Number of items produced=3,840. Standard usage=3,840 × 17·385 lb. =66,758 lb.	
Standard cost=66,758 lb. × standard price of zinc at £100 per ton= £2,980·26½.	Actual cost=70,118 lb. × standard price of zinc at £100 per ton= £3,130·26½.

The material usage variance of £150 unfavourable represents the excess usage of materials (70,118 lb.−66,758 lb.) multiplied by the standard cost of £100 per ton.

Production departments expenses variances

The departmental operating statement for Casting 'A' (Exhibit No. 5) and the schedule of expenses (Exhibit No. 6) show an unfavourable variance on expenses of £232 for the period.

The details of the allowed expenses, £1,694, and of the actual expenses, £1,768, are given on the schedule of expenses. The standard expenses, £1,536, shown in total on the schedule of expenses is equal to the casting department overhead rate of 400 per cent applied to the standard direct labour of the department, £384.

The overhead rate is determined as follows:

Budgeted direct labour of department for year (per production budget, Exhibit No. 18) £4,992

Budgeted expenses for year (per overhead expenses budget, Exhibit No. 19) .. .. £19,968=400 per cent of direct labour.

The schedule of expenses (Exhibit No. 6) shows the full detail of the budgeted expenses for a period of five weeks. Each item is marked as being either fixed or variable or half-fixed and half-variable. The sum of the fixed items amounts to £788 and is equal to 164 per cent of the budgeted direct labour of £480 (£4,992 × $\frac{5}{52}$) and the sum of the variable items amounts to £1,132 and is equal to 236 per cent of the budgeted direct labour.

With the budgeted direct labour of £480 and the standard direct labour of the production achieved of £384 the level of activity is:

$$\frac{384}{480} \times 100 = 80 \text{ per cent}$$

Exhibit No. 24 – continued

	p	
Basic time rate	7·8	per hour
Add Piece-work rate increment at 33– per cent	2·6	
Normal piece-work time rate	10·4	
Revised hourly bonus	7·1	=68 per cent on normal piece-work time rate
Revised standard time rate	17·5	per hour

The revised standard wages are:

	£p
3,840 items produced at 6·25p piece-work rate	240·00
Revised standard uplift for hourly bonus is 68 per cent or 4·25p per item	163·20
=3,840 at 10·50p, revised standard labour rate	£403·20
	say, £403·00

At a revised standard time rate of 17·5p per hour this is equivalent to 2,304 standard hours.

The analysis of total variance is therefore:

(a) *Rates of pay variance (£19 unfavourable)*

	£p
This is the difference between the standard wages of	384·00
and the revised standard wages of	403·20
	£19·20
	say, £19 unfavourable

This represents the increase of 50p in the cost per item of the hourly bonus (from 3·75p to 4·25p) multiplied by the number of items produced, 3,840.

(b) *Efficiency variance (£5 favourable)*

	£p
This is the difference between the revised standard wages of	403·20
and the actual wages of	398·33
	£4·87
	say, £5 favourable.

This represents the decrease of 74 hours, in the actual hours worked (2,230 hours) from the standard hours of production (2,304) hours multiplied by the actual hourly bonus of 7·1p.

Materials usage variance

The departmental operating statement for Casting 'A' (Exhibit No. 5) shows an unfavourable usage variance on materials of £150.

The details of the standard materials, £2,980, and the actual materials, £3,130, are as follows:

THE INTEGRATED ACCOUNTING SYSTEM

Exhibit No. 24

ANALYSIS OF VARIANCES

Price variances

The profit and loss account (Exhibit No. 1) shows a favourable price variance of £278. This is made up of variances arising on each main materials group. For example, the zinc price variance is calculated as follows:

Standard price, £100 per ton. Purchases, 4 tons at £103·50 per ton.

		£p
Standard price		100·00 per ton
Actual price		103·50 ,,
	excess	£3·50 per ton

4 tons × £3·50 per ton = £14·00 unfavourable price variance

Labour variances

The departmental operating statement for Casting 'A' (Exhibit No. 5) shows an unfavourable variance on direct wages of £14.

The details of the standard wages, £384, and the actual wages, £398, are as follows,

Standard Wages		Actual Wages	
	£p		£p
3,840 items produced at 6·25p Piece-work rate	240·00	3,840 items produced at 6·25p Piece-work rate	240·00
Standard uplift for hourly bonus is 60 per cent of 6·25p or 3·75p per item ..	144·00	2,230 hours, time taken at 7·1p actual hourly bonus	158·33
= 3,840 at 10p, standard labour rate	£384·00		£398·33
At a standard time rate of 16½p per hour this is equivalent to 2,304 standard hours		say, £398·00	

The total variance of £13·96 (say, £14) unfavourable is the difference between the standard wages of £384·00
and the actual wages of £397·50

= £13·50 unfavourable

In order to analyse this figure further between rates of pay and efficiency variance it is necessary to recalculate the standard wages on the basis of the actual hourly bonus rate. The standard uplift was established at 6·25p per hour but an increase was made on March 1st to 7·1p per hour. This had the effect of increasing the percentage standard uplift from 60 per cent to 68 per cent as follows:

Exhibit No. 23
facing page 74

Exhibit No. 22 – continued

	Account No.
12. *Transfer of service departments expenditure to production departments, other service departments and chargeable tools and dies work in progress account*	
Dr. Production departments operating (analysed over departments)	60.01
Service departments operating (analysed over departments)	60.21
Chargeable tools and dies work in progress (analysed over labour, materials and tool-room overheads)	07.06
Cr. Service departments operating (analysed over departments)	60.21
13. *Transfer of production departments variances* per departmental operating statements	
Dr. or *Cr.* Labour variances (analysed over production departments)	64.02
Materials variances (analysed over production departments)	64.03
Production department expenses variances (analysed over production departments)	64.04
Dr. or *Cr.* Production departments operating (analysed over production departments) ..	60.01
14. *Transfer of service departments variances*	
Dr. or *Cr.* Service departments expenses variances	64.05
Dr. or *Cr.* Service departments operating ..	60.21
15. *Transfer of standard cost and variances of selling and administration expenses*	
Dr. Standard cost of selling and administration expenses	62.01
Cr. Selling and administration operating ..	60.31
Dr. or *Cr.* Selling and administration expenses variances	64.06

Exhibit No. 22 – continued

		Account No.
8.	**Direct materials issued and charged to departments but not used at end of period** per departmental stock returns	
	Dr. Issued materials suspense (with standard cost of materials unused in production departments)	07.03
	Cr. Production departments operating (analysed over departments)	60.01
	(NOTE. – This entry must be reversed at beginning of the following period)	
9.	**Transfer of direct wages to operating account**	
	Dr. Production departments operating (with amount of direct wages per account 58.01 analysed over departments)	60.01
	Cr. Transfer account	58.99
10.	**Transfer of production and inspection to work in progress** per cost of production summary book	
	Dr. Work in progress (production at standard cost analysed between labour, materials and works overheads)	07.02
	Cr. Production departments operating (analysed over departments including inspection)	60.01
11.	**Transfer of overhead expenditure to operating accounts** per analysed schedule of overhead expenditure	
	Dr. Production departments operating (analysed over departments)	60.01
	Service departments operating (analysed over departments)	60.21
	Selling and administration operating ..	60.31
	Cr. Transfer account (with total of indirect wages and expenses accounts 58.02/72)	58.99

F*   M.A.P.

Exhibit No. 22 – continued

Account No.

4. *Transfer of purchases of materials to stock and work in progress* per purchases day-book
    Dr. Stock of raw materials (with standard cost of purchases in account 54.01) 07.01
    Work in progress (with standard cost of purchases in account 54.02) 07.02
    Cr. Transfer account (with actual cost of purchases in accounts 54.01/2) .. 54.99
    Dr. or Cr. Price Variances (analysed over materials groups) 64.01

5. *Issues of materials from stock* per materials requisitions
    Dr. Direct materials (with issues at standard prices analysed over production departments) 56.01
    Tool-room steel (with issues at standard prices) 56.02
    Cr. Stock of raw materials 07.01

6. *Scrap production* per scrap reports
    Dr. Scrap production (with standard cost of scrap analysed over departments) .. 56.03
    Cr. Work in progress (analysed between labour, materials and works overheads) 07.02

7. *Transfer of materials issued and scrap made to departments operating accounts*
    Dr. Production departments operating (with direct materials per account 56.01 and scrap per account 56.03 analysed over departments) 60.01
    Service departments operating (with tool-room steel per account 56.02 allocated to tool-room) 60.21
    Cr. Transfer account 56.99

THE INTEGRATED ACCOUNTING SYSTEM

Exhibit No. 24 shows examples of how the variances on the profit and loss account and the operating statements are analysed in greater detail by the cost accountant, either on his own initiative or by request. In the initial stages of a management accounting scheme it is often necessary to make such analyses so that foremen and other members of the management can understand more easily the nature of the variances.

Continued on page 81

Exhibit No. 22

ROUTINE COSTING POSTINGS

Account No.

1. *Cost of sales* per costs of sales analysis
    Dr. Works cost of sales (analysed over product groups) 52.01
    Cr. Work in progress (analysed between labour, materials and works overheads) 07.02

2. *Cost of sales returns* per cost of sales returns analysis
    Dr. Work in progress (with works standard cost of goods returned to work in progress – analysed between labour, materials and works overheads) 07.02
    Scrap production (with works standard cost of goods returned and scrapped – analysed over the departments responsible) .. 56.03
    Cr. Works cost of sales (analysed over product groups) 52.01

3. *Cost of tools and dies charged to customers* per cost of sales analysis
    Dr. Cost of chargeable tools and dies 52.02
    Cr. Chargeable tools and dies work in progress (analysed between labour, materials and tool-room overheads) 07.06

F   M.A.P.

from the balances in the balance sheet section of the general ledger, and the profit and loss account (Exhibit No. 1) from the balances in the profit and loss section.

Variances are shown, however, in more detail than in the general ledger accounts (64.01 to 64.06); these accounts only show the total of the major variances analysed by departments. The further analysis required for the profit and loss account is obtained from various statements, such as the summary of works operating departments (Exhibit No. 7).

Exhibits Nos 4 to 8 which summarize or show the results of individual departments, are prepared from the ledger accounts and the various analyses and summaries supporting the 'costing' journal entries (Exhibit No. 22).

Thus, by building up the statements in this way, the general ledger accounts need only be posted in sufficient detail to enable them to serve as control accounts, and a great deal of detailed posting is saved.

The statement of stocks and work in progress (Exhibit No. 3) is compiled from the various general ledger accounts dealing with stocks and work in progress together with the evaluation of a physical count at the end of the period of finished stocks as described in Chapter III.

Analysis of variances (Exhibit No. 24)
The variances which appear on the 'monthly' management statements show the difference between the actual cost incurred and the standard cost under a number of headings. For example, on the profit and loss account (Exhibit No. 1) the price variance is the difference between the purchases valued at standard prices and their actual cost. On the departmental operating statement (Exhibit No. 5) the materials usage variance is the difference between the standard and actual quantities of material used in production at standard prices.

These variances, therefore, can do little more than indicate whether the position, in respect of the costs they represent, is satisfactory or not. If the amount of the variances is insignificant then, normally, no further examination is made. If, however, the variance is large, either in amount or in relation to the total standard cost, then a further analysis is often necessary to identify more specifically the causes of the variances before corrective action can be taken.

THE INTEGRATED ACCOUNTING SYSTEM 69

tion departments (Posting No. 12). The balance on the account, i.e., the difference between (*a*) and (*b*) above, represents the total of the variances for the service departments for the period. This balance is analysed by departments – there is no breakdown by elements of cost as in the case of the production departments – and is transferred by Posting No. 14 from the service departments operating account to the service departments variances account.

Posting No. 15 – Transfer of the standard cost and variances of selling and administration expenses

This is the final routine 'costing' posting which transfers the standard cost of selling and administration for the period to a separate account in the general ledger. The charge is determined by applying the budgeted overhead rates for selling and administration to the total of the works standard cost of sales. The standard cost is credited to the selling and administration operating account; the actual expenses have already been charged to the account by Posting No. 11, and the difference between standard cost and the actual expenses is transferred by Posting No. 15 from the operating account to the selling and administration variances account.

Illustration of general ledger accounts (Exhibit No. 23)

Postings to the general ledger accounts are made from the books of prime entry and the 'costing' postings shown on Exhibit No. 22. As described earlier in this chapter, these postings are quite straightforward, but to illustrate the entries in the general ledger involved in arriving at the standard cost of sales for the period, the accounts for stock of raw materials, work in progress, production departments operating and cost of sales are shown on Exhibit No. 23. The reference number appearing against an entry is that of the 'costing' posting on Exhibit No. 22.

Preparation of management accounting statements

The management accounting statements are prepared for each accounting period after all the postings have been made to the general ledger and the ledger agreed by listing the balances on an adding machine.

The statement of financial position (Exhibit No. 2) is compiled

charge is used instead of an actual charge for the period because, until a system of flexible budgets is instituted, works general and stores and dispatch expenses are treated as wholly fixed, i.e., they are not considered to vary with the level of production. This means that no volume variance is shown on the operating statements for these departments, and the difference between the budgeted and actual expenses is shown as an expense variance.

The maintenance department charges the production departments and the other service departments with the wages recorded as being incurred on work for these departments uplifted by the overhead rate for the maintenance department. The charge made by the tool-room to other departments is similarly calculated. This posting also transfers to a special work in progress account the material, labour and overhead cost of making tools and dies to be charged to customers.

Posting No. 13 – Transfer of production departments variances
The production departments operating account (analysed over departments) now contains (*a*) the charges for the actual expenditure, i.e., raw materials at standard prices (Posting No. 7), direct labour (Posting No. 9) and overhead expenses (Postings Nos 11 and 12); and (*b*) the credit for the standard cost of production analysed by departments over the same elements of cost (Posting No. 10). The balance on the account, i.e., the difference between (*a*) and (*b*) above, will represent the total of all the variances incurred in the production departments for the period; this balance is then analysed to obtain the material usage, labour and expenses variances for each department. Posting No. 13 then transfers these variances from the production departments operating account to the labour variances account, the materials variances account and the production departments expenses variances account.

Posting No. 14 – Transfer of service departments variances
The service departments operating account contains postings analysed to departments for (*a*) charges for actual expenditure in respect of tool-room materials at standard prices (Posting No. 7), wages and expenses (Posting No. 11) and transfers from other service departments (Posting No. 12); and (*b*) credits for services rendered to other service departments and the produc-

by each department is then credited to its operating account and the work in progress account is debited.

The production reports which deal with products passing to finished stocks are also used to evaluate the standard cost of inspection. For each product the standard cost of inspection is taken from the product standard cost card. The total standard cost for all products is then debited by this posting to the work in progress account and credited to the inspection department's operating account.

Posting No. 11 – Transfer of overhead expenditure to operating accounts

Once the financial postings are completed, the monthly totals appearing in the indirect wages and expenses accounts (58.02 to 58.72) in the general ledger are entered on a schedule of overhead expenses.

This schedule is drawn up in the same form as the overhead expenses budget (Exhibit No. 19) with columns for the departments across the page, and each item of expense is allocated or apportioned to the departments on the same basis. The schedule thus shows the detailed expenses for each department.

Posting No. 11 is then prepared; this posting charges the expenses to the operating accounts of the various production and service departments.

The total expenses are credited to the transfer account (58.99); with this posting and Posting No. 9 this account now contains a credit representing the whole of the expenditure in the group of accounts for wages and expenses. This group of accounts is now self-balancing and the wages and expenses accounts themselves are not encumbered with the transfers to the departmental operating accounts, thus facilitating the preparation of the conventional form of profit and loss account at the end of the financial year.

Posting No. 12 – Transfer of service departments expenses

This posting transfers from the various service departments their charges for services rendered to other service departments and the production departments.

The amounts charged to these departments by works general and the stores and dispatch departments are the proportion for the period of the budgeted charge for the year. A budgeted

Posting No. 7 – Transfer of materials issued and scrap made to departments operating accounts

This posting charges the production departments and the tool-room with the materials they have used in the month, and the scrap for which they are responsible, as stated in Postings Nos 5 and 6; a corresponding credit is made to the transfer account (56.99). This transfer account enables the totals appearing in the accounts in group 56 (materials used from stock and scrap made) to be transferred to the departments operating accounts without crediting the individual accounts. Group 56 is therefore self-balancing and is not needed when preparing a profit and loss account in standard cost form; neither are the accounts in this group required for a conventional form of profit and loss account as the figures in the previous group (purchases of direct materials) would be used.

Posting No. 8 – Direct materials issued and charged to jobs but not used at end of period

The purpose of this posting is to transfer to a suspense account the standard cost of the raw materials issued and charged to departments but unused at the end of the period. This involves a small amount of stock-taking. The entry is reversed in the following accounting period.

Posting No. 9 – Transfer of direct wages to operating account

As in the case of direct materials, a transfer account is again credited when direct wages are transferred to the production departments operating accounts. Wages are analysed weekly over all departments and between direct wages, indirect wages, holiday credits and National Insurance, and a 'financial' posting is made to the various wages accounts (58.01 to 58.04) in the general ledger.

Posting No. 10 – Transfer of production and inspection to work in progress

Production reports are received weekly from the progress department. The cost office summarizes the quantities leaving each department under product or part numbers, and evaluates them at works standard cost by reference to the product standard cost cards. The total standard cost of the work done

The total cost of tools and dies completed and charged to customers during the period is debited by this posting to the chargeable tools and dies account and credited to the work in progress account.

Posting No. 4 – Transfer of purchases of materials to stock and work in progress

This posting transfers (*a*) the purchases of raw materials to the stock account and the purchases of bought-out parts to the work in progress account at standard prices; and (*b*) the differences between standard and actual prices to the price variance accounts. Bought-out parts are posted direct to the work in progress account and are therefore not charged to the department when drawing the parts from store; this reduces the clerical work in evaluating stores requisitions. Price variances are transferred to price variance accounts at the purchase invoice stage and materials and bought-out parts are dealt with in the cost accounts at standard prices. A column in the purchases analysis book is used to record the value of each invoice for materials and bought-out parts at standard prices. A material classification code is used to analyse the totals of the purchases and price variances over material groups.

The actual value of purchases is debited to the purchases of raw materials and bought-out parts accounts and credited to the purchases ledger control account by a financial posting from the purchases day-book.

Posting No. 5 – Issues of materials from stock

By this posting the standard cost of raw materials drawn from store in the month is debited to the direct materials account (analysed over the using departments) and the tool-room steel account, and credited to the stock of raw materials account.

Posting No. 6 – Scrap production

The standard works cost of scrapped work for the month is debited in this posting to the scrap production account (analysed over the responsible departments) and credited to the work in progress account. This is necessary because Posting No. 10 includes the cost of scrapped work in the standard cost of production which is debited to the work in progress account.

A prepayments and accruals ledger, as illustrated in Exhibit No. 21, is used. This ensures that expenditure is charged to the general ledger accounts in the correct period either by accruing expenditure for which a charge has not yet been received, or by carrying forward prepaid expenditure. The general ledger is entirely self-balancing.

Once the 'financial' postings had been made, and if the figure of closing stock and work in progress were known, a set of accounts in traditional form could be prepared showing the profit for the period. As the system of standard costing provides a cost-of-sales figure, the profit for the period can be determined without any reference to closing figures of stock and work in progress.

The monthly postings based on the costing records are first entered in a cost journal. This journal is quite distinct from the financial journal and is written up by the staff of the cost office. Nevertheless, the entries do indirectly affect the financial accounts in view of their bearing on the valuation of stock and work in progress.

The routine monthly postings based on costing records are shown in Exhibit No. 22 (see page 71), and they are explained below:

Posting No. 1 – Cost of sales
This represents the total of all the sales invoices valued at works standard cost as shown by the product standard cost cards. The total works standard cost for each product group is analysed between raw materials, direct labour and works overheads to correspond with the analysis of the work in progress account.

Posting No. 2 – Cost of sales returns
This posting debits the work in progress account with the works standard cost of goods returned and put back into work in progress for rectification. If returned goods are scrapped, however, the works standard cost is debited to the scrap production account and analysed to the department responsible for the defects.

Posting No. 3 – Cost of tools and dies charged to customers
Job records are kept of the labour, materials and overhead cost of the tools and dies made for, and charged to, customers.

Exhibit No. 21

PREPAYMENTS AND ACCRUALS LEDGER
Illustration of Account

Light, Heat and Power

No. 58.13

Date	Narrative	Ref.	Dr.	Cr.	Electric light	Electric power	Gas	Coke
1972			£	£	£	£	£	£
Jan. 1st	By Balances accrued	b/f.		105 00	7 00	38 00	60 00	
	To Stock of coke	b/f.	45 00					45 00
Period 1	By Transfer to expense account per measured consumption	J.		80 00	5 00	25 00	40 00	10 00
	To Purchases	J.	90 00		6 00	34 00	50 00	
	To Balance	c/d.	50 00					
			185 00	185 00				
	By Balance	b/d.		£50 00	£6 00	£29 00	£50 00	£35 00

Exhibit No. 20 – continued

Group No.	Group	Account No.	Account
60	Operating accounts	01	Production departments operating (analysed over departments)
		21	Service departments operating (analysed over departments)
		31	Selling and administration operating
62	Standard cost of selling and administration expenses	01	Standard cost of selling and administration expenses
64	Variances	01	Price variances (analysed over various headings)
		02	Labour variances (analysed over production departments)
		03	Materials variances (analysed over production departments)
		04	Production departments expenses variances (analysed over production departments)
		05	Service departments expenses variances (analysed over service departments)
		06	Selling and administration expenses variances
66	Miscellaneous income	01	Discounts received
68	Miscellaneous expenditure	01	Loss on sale of fixed assets
69	Taxation	01	Income tax, Schedule D
70	Appropriations	01	Dividend

THE INTEGRATED ACCOUNTING SYSTEM

Exhibit No. 20 – continued

Group No.	Group	Account No.	Account
58	Wages and expenses – contd	32	Carriage inwards
		33	Carriage outwards
		34	General packing materials
		35	Agents' commission
		36	Motor-car running expenses and repairs
		37	Motor van running expenses and repairs
		38	Discounts allowed
		41	Advertising and exhibition expenses
		42	Printing and stationery
		43	Postage and telephones
		44	Rental of broadcasting equipment
		45	Sundry office expenses
		46	Travelling and entertaining expenses
		47	Welfare
		48	Canteen expenses
		61	Bad debts
		62	Night security
		63	B.S.I. and analytical charges
		64	Hire-purchase interest
		65	Bank interest
		66	Bank charges
		67	Legal and professional expenses
		68	Audit fee
		71	Directors' remuneration
		72	Depreciation (analysed over asset accounts)
		99	Transfer account (to operating accounts)

Exhibit No. 20 – continued

Group No.	Group	Account No.	Account
58	Wages and expenses	01	Direct wages (analysed over departments)
		02	Indirect wages (analysed over departments)
		03	Holiday credits (analysed over departments)
		04	National Insurance on wages (analysed over departments)
		05	Salaries (analysed over departments)
		06	National Insurance on salaries (analysed over departments)
		07	Group life and pension scheme Analysed over: (*a*) Life assurance (*b*) Past pension (*c*) Company's current contributions
		11	Rent
		12	Rates
		13	Light, heat and power (analysed over items of expense)
		21	Production departments general expenses (analysed over production departments)
		22	Service departments general expenses (analysed over service departments)
		23	Factory general expenses
		24	Consumable tools
		25	Repairs and replacements to plant, machinery and equipment (analysed over departments)
		26	Repairs to premises
		31	Insurances (analysed)

THE INTEGRATED ACCOUNTING SYSTEM

Exhibit No. 20 – continued

Group No.	Group	Account No.	Account
13	Shares in subsidiaries	01	A.B. Co Ltd
		02	C.D. Co Ltd
15	Current accounts with subsidiaries	01	A.B. Co Ltd
		02	C.D. Co Ltd
19	Share capital	01	Ordinary shares
21	Capital reserves	01	Capital reserve
23	Revenue reserves and surplus	01	Profit and loss account
		02	Reserve for future income tax

Profit and Loss Account Section

Group No.	Group	Account No.	Account
50	Sales	01	Sales (analysed over product groups A., B., C., etc.)
		02	Chargeable tools and dies
		03	Sales of scrap
52	Works cost of sales	01	Works cost of sales (analysed over products groups A., B., C., etc.)
		02	Cost of chargeable tools and dies
54	Purchase of direct materials	01	Raw materials (analysed over various headings)
		02	Bought-out parts
		99	Transfer account (to stock, variance and work in progress account)
56	Materials used from stock and scrap made	01	(Analysed direct materials over departments)
		02	Tool-room steel
		03	Scrap production (analysed over departments)
		99	Transfer account (to operating accounts)

Exhibit No. 20 – continued

Group No.	Group	Account No.	Account
07	Current assets – contd	21	Sales ledger control
		22	Provision for doubtful debts
		23	Provision for discounts
		25	Prepaid expenses control
		26	Cash at bank control – No. 1 Account
		27	Petty cash control
		29	Employees' loan accounts
		30	Group life and pension scheme premiums – Analysed over: (a) Life assurance (b) Past pension (c) Company's current contributions (d) Employees' current contributions
		31	Pension deductions from employees
09	Current liabilities	01	Purchases ledger control
		02	Creditors under hire-purchase accounts
		03	Accrued expenses control
		04	Wages and National Insurance control
		05	Salaries and National Insurance control
		06	P.A.Y.E.
		07	Holiday credits control (Separate account for each year, July 1st–June 30th)
		08	Current taxation
11	Bank loans	01	Overdraft at bank control – No. 2 Account

THE INTEGRATED ACCOUNTING SYSTEM

Exhibit No. 20

CODE OF ACCOUNTS

BALANCE SHEET SECTION

Group No.	Group	Account No.	Account
01	Fixed assets	01	Leasehold land and buildings
		03	Building extension
		05	Plant and machinery
		07	Fixtures and fittings
		09	Office furniture
		11	Commercial vehicles
		13	Motor-cars
		15	Canteen equipment
		17	Loose tools
		19	Plant and machinery etc. on hire-purchase
		50	Capital work in progress account
03	Depreciation provisions	01 etc.	Leasehold land and buildings (as fixed assets)
07	Current assets	01	Stock of raw materials
		02	Work in progress (analysed over materials, labour and works overheads)
		03	Issued materials suspense
		04	Stock and work in progress reserve
		05	Stock and work in progress adjustment
		06	Chargeable tools and dies work in progress (analysed over materials, labour and tool-room overheads)
		07	Packing materials stock
		08	Consumable stores stock
		09	Consumable tools stock
		10	Printing and stationery stock
		11	Replacements to plant, machinery and equipment stock

Chapter VIII

THE INTEGRATED ACCOUNTING SYSTEM

The costing and financial accounting system of the company is described in this chapter.

General ledger
One general ledger is used, divided into two sections comprising:
   (a) those accounts which are shown in the balance sheet; and
   (b) those accounts which are transferred to the profit and loss account.

Each section is subdivided into groups of accounts of a like nature, e.g., fixed assets, current liabilities, expenses. In the profit and loss section there are, in addition to the normal revenue and expenditure accounts, a number of accounts which deal exclusively with costing matters, e.g., cost of sales, transfer and variance accounts. When producing a traditional profit and loss account at the end of the financial year, these 'costing' accounts are ignored but they are used when producing the monthly profit and loss accounts in standard cost form.

Each account in the general ledger is given a four-figure code number, the first two figures indicate the group and the last two figures indicate the account within the group. The code of accounts used in this particular system is illustrated in Exhibit No. 20.

Postings during each accounting period
The company's accounting period consists of four or five weeks, two four-week periods being followed by one of five weeks. These periods are rather loosely referred to as 'months'. The financial books of prime entry are posted to the general ledger in the usual way. As the financial accounting system is quite simple the postings are generally made direct from the daybooks; only special items are posted through the medium of the journal. Control accounts are maintained for debtors, creditors, prepayments, accruals and cash.

Continued on page 64

STANDARD COSTS AND BUDGETS

is then expressed as a percentage of direct labour for the group as shown by the production forecast.

Finally, the totals of the expenses for administration and selling are converted into overhead rates expressed as percentages of the total works standard cost of budgeted sales. For example, the administration overhead rate of 10·24 per cent is calculated as follows:

                                                                         £

Budgeted administration expenses per Exhibit No. 19 .. 43,598
Works standard cost of budgeted sales per Exhibit No. 17
   excluding chargeable tools 425,792

$$\frac{43{,}598}{425{,}792} \times 100 = 10\cdot24 \text{ per cent}$$

Along with the other forecasts the overhead expenses forecast is considered by the board and, when approved, becomes the overhead expenses budget. This budget serves two distinct functions; it provides a means of control by comparing actual overhead expenses with the budgeted expenses and examining the differences or variances, and it provides departmental overhead rates which are entered on the product standard cost cards (Exhibit No. 16).

Although the accountant takes the initiative and is responsible for organizing the preparation of the budgets, he works very closely with the works manager, departmental managers and foremen who are largely concerned with incurring expenditure.

Forecast and budgeted profit and loss account

Once the sales forecast, production forecast and overhead expenses forecast have been prepared they are brought together in a forecast profit and loss account. The form of the profit and loss account follows the lines of Exhibit No. 1 without, of course, the variance items. The amount of the profit thus disclosed is a major factor which the board takes into account when considering the budgets for the year. The board also considers a forecast statement of the financial position at the end of the financial year and a cash forecast for the year.

The expenses of the stores and dispatch department are apportioned on the basis of direct wages of production departments, plus the wages of the tool-room and the maintenance department. The amount of £388 apportioned to Casting 'A' department is calculated as follows:

	Total	Casting 'A'
Stores and dispatch expenses		£12,432
Budgeted wages:	£	£
Standard labour cost (direct wages)	112,800	4,992
Tool-room wages	39,200	
Maintenance department wages	7,950	
	£159,950	£4,992

$$\frac{4,992}{159,950} \times 12,432 = £388$$

The expenses of the maintenance department are then apportioned over all production departments and the tool-room on the basis of estimated usage of the maintenance facilities, using the analysis of labour over past periods as a guide.

Finally, the tool-room expenses are allocated between (*a*) tools which will be recharged to customers and (*b*) tool-room services for production departments, the total cost being apportioned over these departments in the same way as the expenses of the maintenance department.

The total of the expenses of each production department (with the exception of the inspection department) is expressed as a percentage of the standard labour cost of the department to produce the departmental overhead rate. These departmental overhead rates are based on the forecasted level of production as represented by the standard labour cost in the production forecast. It will be seen, therefore, that where the forecasted level of production of a department is below its productive capacity, the cost of the unused capacity is absorbed in the overhead costs of all products passing through the department.

Next, the expenses of the inspection department are apportioned to product groups. This apportionment is based on an estimate of how the inspectors will spend their time on each group of products, taking into account an actual analysis of past time and the types of products to be produced during the year. The total expense apportioned to each product group

expenses, as the name implies, vary with but not in direct ratio to the volume of production. It was decided to treat these semi-variable expenses as wholly fixed or wholly variable or, in some cases, to consider them as partly fixed and partly variable. At a later stage in the development of the scheme it may be intended to introduce flexible budgets, and variable and semi-variable expenses will then be budgeted for various levels of production; this development is explained in some detail in Chapter IX.

The overhead expenses are then entered on the overhead expenses forecast, item by item. Some expenses are estimated by the accountant in total and then apportioned over the departments, e.g., rent and rates are apportioned on the basis of floor space occupied by each department. Other expenses, e.g., indirect wages, scrap and consumable tools and stores, are estimated for each department by the cost accountant in co-operation with the works manager. Where expenses such as salaries can be identified with a particular department, they are allocated to that department by the accountant.

Each department is charged with scrap at the full works standard cost, i.e., materials, direct labour and works overheads up to the point of rejection.

The expenses have now been allocated or apportioned over all the departments and we come to the work of apportioning the total cost of each service department to the production departments and also, in some cases, to other service departments, since one service department may carry out work for another service department.

The works general expenses are apportioned on the basis of total budgeted wages for all departments, including other service departments. The amount of £1,218 apportioned to Casting 'A' department is calculated as follows:

Works general expenses			£33,270
		Total	Casting 'A'
Budgeted wages:		£	£
Standard labour cost (direct wages)		112,800	4,992
Indirect wages:			
Waiting time		2,050	280
Other indirect labour		99,028	2,558
		£213,878	£7,830

$$\frac{7,830}{213,878} \times 33,270 = £1,218$$

Acc. code	It...
58.01	Stand... pro...
58.02	Indire... Wa... Oth...
.03	Holid...
.04	Nat. ...
.05/.06	Salari...
.07	Grou...
11/12	Rent ...
13	Lighti...
,,	Powe... Etc.
56.03	Scrap...
	Tot...
	Reap... *vic...* Work... Store... Main... Tool-...

Total departm...

Departmental...

Exhibit No. 19
facing page 52

analysis over elements of cost can be made; no estimate of quantities is possible.

An adjustment is then made to allow for the building up or running down of work in progress and finished stocks planned during the year. As will be seen from Exhibit No. 18, the adjustment is expressed as a percentage of the prime standard cost of the forecasted sales for each product group. It was not considered necessary or practicable in the early stages of the scheme to make this adjustment for each product or to calculate it more meticulously. The forecasted production for each product group is then summarized to show the total raw materials and labour costs for each production department.

The production forecast is then considered by the works manager in the light of his available productive capacity and labour force. It is quite reasonable in this particular case to use the estimated cost of material and labour as a basis for a preliminary comparison between what is required and what is available by way of plant and labour. If any fundamental decisions had to be taken a more accurate assessment of capacity and requirements would be made.

Should the forecasted requirements be in excess of the capacity of the factory departments, decisions are taken, if necessary, at board level, either to adjust the quantities in the sales forecast or to provide additional productive facilities. Finally, the production forecast is agreed by the board and becomes the production budget as illustrated in Exhibit No. 18.

Overhead expenses forecast and budget (Exhibit No. 19)
An overhead expenses forecast, on which the overhead expenses budget is based, is prepared by the accountant and the cost accountant in conjunction with the works manager, in the light of the departmental production figures given in the production forecast.

Items of expense are first listed on the forecast in the same order as they appear in the general ledger. There is a column showing whether the expense is fixed or variable in relation to production. Fixed expenses are those which are considered not to fluctuate with the volume of production, e.g., rent, rates, salaries. Variable expenses are those expenses which are considered to vary directly with production, e.g., power, scrap.

Other expenses usually referred to as semi-variable expenses are neither fixed nor directly variable with production; these

STANDARD COSTS AND BUDGETS 51

ments. Since the description and detailed specification will not be known until the customers' inquiries are received, only the value of these sales can be forecasted; quantity and prices cannot be stated. Repeat orders from customers can, however, be forecasted in the same way as the company's own products, that is in terms of quantity, price and value.

When the sales forecast has been reconciled with the other forecasts and has been agreed by the board, it becomes the sales budget of the company. Until then the figures are only provisional. The sales manager may be asked, for instance, to revise his sales forecast for certain products because there is not enough productive capacity in some departments to meet his forecast. In this event consideration may have to be given to the possibility of increasing prices in order to offset, at any rate to some extent, the loss of volume; alternatively, sales effort may have to be concentrated on other products.

The company's sales budget is illustrated in Exhibit No. 17. In addition to the information included in the sales forecast as mentioned above, the sales budget shows the works standard cost and the standard gross margin; the standard gross margin is the difference between the standard selling price and the works standard cost of each product and is the contribution to administration and selling overheads and profit. These margins are worked out from the information shown on the product standard cost cards, and are used as a guide to the relative profitability of the products when considering such problems as the alternative use of productive capacity and the concentration of sales effort.

Production forecast and budget (Exhibit No. 18)
A production forecast is then prepared.

The quantities of each product given by the sales forecast are first entered on the production forecast. These forecasted sales are then evaluated in terms of prime standard cost (i.e. standard cost of direct materials and direct labour) as shown on the product standard cost cards. The prime standard cost for each product is broken down into (*a*) bought-out parts and (*b*) raw materials and direct labour analysed over production departments.

The last item in the production forecast is new lines, for which only an estimate of the prime standard cost and its

Item No.	Quantity per sales budget	P
1	4,000	
2	3,000	
3	8,000	
etc.		
New lines		

Plus increase in progress and stocks – 7·5%

Pro

Product group	Pr stan c
A ⎫	4(
B ⎬	18:
C ⎪	
etc. ⎭	
	£229

Exhibit No. 17

SALES BUDGET
Fifty-two weeks ended December 29th, 1972
PRODUCT GROUP BUDGET – DETAILED

PRODUCT GROUP A

Item No.	Quantity	Standard Sales Value		Works Standard Cost		Standard Gross Margin		
		Per Unit	Amount	Per Unit	Amount	Per Unit	Amount	% on Sales
		£p	£	£p	£	£p	£	
1	4,000	1·00	4,000	0·642	2,569	0·358	1,431	
2	3,000	1·75	5,250	1·043	3,125	0·707	2,125	
3 etc.	8,000	0·75	6,000	0·375	3,000	0·375	3,000	
			80,558		41,128		39,430	
New lines			50,000		26,000		24,000	
			£145,808		£75,822		£69,986	48·0%

PRODUCT GROUP BUDGET – SUMMARY

Product group	Standard sales value	Works standard cost	Standard gross margin	
	£	£	£	% on Sales
A	145,808	75,822	69,986	48·0
B	79,248	39,624	39,624	50·0
C	114,402	67,496	46,906	41·0
D	84,762	63,566	21,196	25·0
E	193,254	171,996	21,258	11·0
F	8,996	7,288	1,708	19·0
	626,470	425,792	200,678	32·0
Chargeable tools	14,332	9,416	4,916	34·3
	£640,802	£435,208	£205,594	32·1%

Inspection costs are shown after all the other manufacturing costs have been recorded and are not, therefore, included in cost of work in progress until the final assembly point. It would be more accurate to include inspection costs after each operation but the clerical work involved would be prohibitive.

The stage has now been reached where the works standard cost of the product is shown on the product standard cost card. The total standard cost of the product is arrived at by applying an administrative overhead rate and a selling overhead rate to the works standard cost of each product. These rates have been arrived at by expressing the budgeted administration and selling expenses as percentages of the works standard cost of budgeted sales. The resulting overhead rates are noted on the overhead expenses budget, Exhibit No. 19.

The product standard cost card is then completed by entering the standard selling price, the standard gross margin on works standard cost, and the analysis of the standard gross margin. This information is not usually shown on a product standard cost card but it has been found useful here as the card will be used later, as described in Chapter IX, to evaluate sales invoices.

Forecasts and budgets

Towards the end of the year the work of preparing the following year's forecasts and budgets begins. The bulk of the work is done by the accountant and cost accountant in association with other executives who they ask for information about such matters as sales quantities and prices, production costs and, especially, capital expenditure.

In this section I refer to forecasts and budgets. In this exercise, a forecast is a provisional estimate of what will take place in a certain period; a budget is an agreed estimate, approved by the board and accepted as the plan for the particular period.

Sales forecast and budget (Exhibit No. 17)

A sales forecast is first prepared. In this forecast the number, the sales price and sales value of the company's products which the sales manager expects to sell are listed under each product group.

The sales forecast also includes an estimate of the sales of new products which will be made to customers' own require-

passes and the operations performed are entered from the product specification prepared by the drawing office. The product specification also provides the information for entering the quantity of raw material, including an allowance for waste and scrap required in the casting, and the part to be acquired from outside.

When the details of the product specification are entered on the product standard cost card, the cost office staff can record the costing information. The standard prices of raw materials and bought-out parts are entered in the price per unit column and the amounts are extended to the appropriate columns. The standard labour rate is entered in the direct labour column.

The next stage in the preparation of the product standard cost card is to calculate and enter the amount of works overhead cost to be charged to the product.

Before this can be done, however, the overhead expenses and direct wages for each production department have to be budgeted. The overhead expenses are then expressed as a percentage of the direct wages and the appropriate departmental rate is applied to the standard labour cost of each operation appearing on the standard product cost card. The overhead rates used on the product standard cost card have been taken straight from the overhead expenses budget, an abridged version of which is shown as Exhibit No. 19.

The cost of each operation and the total cost after each operation can now be entered in the work in progress column.

The next item to be dealt with is cost of inspection. Work is generally counted and inspected after each operation but inspection is not necessarily 100 per cent. For example, in the case of simple castings inspection may consist of a sample examination of a certain number of items in a batch; the inspection of scientific instruments is, however, 100 per cent and includes very rigorous tests.

As, therefore, the degree of inspection depends on the product, the cost of inspection is dealt with as a separate item on the product standard cost card. An inspection department rate based on total direct labour cost is established for each product group, as explained later in this chapter. The appropriate rate is then applied to the total direct labour cost of the product as shown by the product standard cost card.

STANDARD COSTS AND BUDGETS

Exhibit No. 16

PRODUCT STANDARD COST CARD

Product Group: A   Item No.: 1   Description: Zinc – Animal

Department operation and materials	Price per unit	Bought-out parts	Raw materials	Direct labour	Overheads	Work in progress Dept	Work in progress Cum.
	p	p	p	p	%	p	p	p
Casting 'A'								
Casting – 1 lb. zinc	39		39·00				39·00	
labour ..				6·50	400·00	26·00	32·50	
							71·50	71·50
Machine Shop								
Turning				14·00	130·00	18·20	32·20	103·70
Assembly No. 1								
Assembly with base								
1 base	27	27·00					27·00	
labour				4·00	118·83	4·75	8·75	
							35·75	139·45
Prime Standard Cost								
90·50p		27·00	39·00	24·50				
Inspection								
60 per cent of direct								
labour						14·70	14·70	
Works Standard Cost		27·00p	39·00p	24·50p		63·65p		154·15
Administration								
10·24 per cent on works standard cost							15·78	
Selling								
6·14 per cent on works standard cost							9·46	
							25·24	
Total Standard Cost								179·39p
Standard Selling Price (per Sales Budget) (240p)								240·00p
Standard Gross Margin on Works Standard Cost								85·85p
Analysis of Standard Gross Margin:								
Administration and selling overheads							25·24	
Leaving a standard net profit of							60·61	
							85·85p	

merit of being one of the simplest to apply and, in any case, raw materials are not the major item of cost in their products.

Standard labour rates

The method of determining the standard labour rates to be used in a standard cost scheme depends, to a large extent, on the particular system of wage payments in operation.

In this company a piece-work system operates for most of the direct labour. The piece-work rates once fixed are not revised unless the methods used on the job are changed. A small volume of work, of a non-recurring nature, and jobs for which satisfactory piece-work rates cannot be determined at the start, are remunerated at time rates according to the class of worker. Direct operatives when on waiting time, cleaning, and tea breaks, etc., are paid at time rates. In addition to the piece-work rates and time rates, an hourly bonus is paid for all hours worked whether on direct or indirect work, and it is this hourly bonus which is adjusted when wage awards are granted.

To make the work of setting standard labour rates as simple as possible, every operation, whether normally remunerated on a piece-work or a time basis, is given a standard labour rate. In the case of a piece-work job, the piece-work rate is taken and is uplifted by a percentage to cover the hourly bonus rate as at January 1st of each year. In the case of time work a corresponding piece-work rate is calculated on the assumption that the job takes a certain time as assessed by the rate-fixers; the resulting piece-work rate is then uplifted by the percentage to cover the hourly bonus rate.

Product standard costs

A product standard cost card is made out for each product manufactured. Where a product consists of a number of parts, either made in the factory or bought out, a standard cost card is prepared for each of the factory-made parts. A summary card is then prepared for each sub-assembly and final assembly; the bought-out parts are dealt with in the appropriate sub-assembly or final assembly cards.

A product standard cost card is illustrated in Exhibit No. 16; this is for a product consisting of a zinc casting made in the factory which after machining is assembled with a part bought from outside. The departments through which the product

CHAPTER VII

STANDARD COSTS AND BUDGETS

This chapter describes how standard costs and budgets are prepared for the concern whose management accounting statements have already been explained in Chapter III. It will be recalled that this concern employs about three hundred people and produces mechanical and scientific instruments as well as castings.

The system of management accounts is based on a scheme of standard costs and budgetary control which enables actual results to be compared with budgets for many significant aspects and divisions of the company's activities. The scheme is integrated with the normal financial accounts through the medium of the general ledger; this integration ensures that the financial and the cost accounts always agree one with the other.

Standard material prices
Standard prices are fixed for all raw materials and bought-out parts. The main items are raw materials for castings and mouldings, and steel and brass for machined parts and tools. The bought-out parts consist of a number of minor items required for mechanical and scientific instruments. The standard prices are taken to be the probable market prices ruling at January 1st; they are fixed by the buyer and confirmed by the managing director. For any new item purchased during the year, which was not included in the production budget and therefore for which a standard price had not been fixed, the standard price is fixed at the price first paid.

There are other methods of determining standard prices; for example, they could be based on the estimated prices to be paid half-way through the financial year. Some companies revise their standard prices for every substantial change in the purchase price of their larger items of raw materials. Each basis has its merits but this company selected the probable market price at the beginning of the year and no further adjustment is made throughout the year. This basis has the

report to the management that all targets have been met precisely and no accounting returns are necessary.

In practice this is unlikely to happen as at all times movements for the better or worse are taking place. These variations from the plan are the signals for corrective action and may be signs of emerging trends to avoid or to take advantage of.

This method clarifies and simplifies the manager's problems and leads to greater clarity.

The misuse of management accounting
Management accounting as a technique has a very sharp cutting edge and as with all such tools can be dangerous.

Properly used, management accounting can permit management to give greater delegation, by targets create incentives, by plans in financial terms give navigational aid to those responsible for control areas.

If, however, management regards such accounts, not as a means for greater freedom and greater initiative but as a means for restraint and punitive measures, then the accounting system can be a very demoralizing influence.

The management accountant's attitude is all important. If reports are continuously critical of all unsatisfactory variations from standard, if management is encouraged to take up a critical viewpoint, then the management accounting system will fail to be a welding and encouraging influence.

The accountant would be looking for opportunities for company operational improvement, to make the most of the skills and strengths of members of the organization and encourage initiative at all levels.

Although as mentioned previously, the management accountant as such has no executive authority outside his own department, in fact by the full use of his advisory function and by giving a first-class information service designed to meet management's needs at all levels, he can occupy a position of great and constructive influence in the organization.

Reviewing standards and budgets

A comparison with last year's figures is usually valueless, as the situation is always changing; so it is with standards and budgets.

Standards and budgets should therefore be revised at suitable intervals.

Budgets are usually reviewed annually, although these may be adjusted in the interim when major changes occur such as a large national wage award.

Some concerns where trade is seasonal or subject to considerable fluctuation, prefer to use budgets of short duration – for example, three months. The largest company in the world uses a method of rolling budgets so that each month it is up-dated.

There are, of course, short-term and long-term budgets, the latter running up to periods of five years. The principle is, the shorter the period, the more precise the figures should be.

Standards are of different nature. Whereas budgets are usually effected by outside factors, standards are usually internal concepts changing only as internal methods and action arises. Examples are: time allowances for operation which change as machines and tooling changes, the effective use of materials, scrap, etc.

Standards should be subject to continual scrutiny to obtain greater efficiency and reduced costs and these should be shown up in the standard cost allowances and profits.

Management by exception

In control by figures, the problem is always the overcoming of excessive figures and information. Computers can be a curse in this respect.

Therefore the best precept is that of control by examining only the movements away from the plan.

In theory, and in a perfect situation, it should be possible to

(f) the variances which are significant to management, in each individual case a selection has to be made of those variances which are desirable and which it is practical to produce without undue cost in labour and paper;

(g) the principles of organization, to ensure that his management accounts are based on a sound organization plan.

Accurate basic information
The effectiveness or otherwise of all management accounts depends on the accuracy of the data from which the figures are prepared; this basic data takes many forms and much of it originates outside the accounts department. Once the right figures have been established, the preparation of the ultimate accounts is largely a matter of routine, even if of a somewhat specialized nature.

It is essential, therefore, for the management accountant to be able to judge whether all the data flowing into the accounts is accurate and satisfactory. Before he is satisfied that this is so, changes will nearly always be necessary to the records and systems of the production, service, sales and administration departments which originate the data; the accountant must be prepared to advise on these changes and to see that the system as revised is within the capacity and capabilities of the staff of the departments concerned.

Communications
The management accountant's work in collecting and processing information is of little interest to management provided it is economically and efficiently carried out.

The accountant's skill arises in selecting and disseminating the information. It is in reporting and interpreting figures that his work is recognized. The better the accountant's knowledge of the business the better the interpretation he can make upon the figures he produces.

It is seldom sufficient to distribute accounting reports and statements and leave it at that.

A follow-up with personal contact in which the figures and possibilities of action required is discussed, will lead to a dynamism in the organization arising from effective use of the accounting information.

that the minimal profit cannot be obtained, then failing a fresh plan to give adequate return, the question of cessation of business in that form must arise.

It must be clearly understood that it is not the management accountant's responsibility to prepare a plan. It is wholly the responsibility of management; the accountant supplying information, recording, measuring, and finally summarizing.

The first step is the master plan agreed by the top management, based on sales figures, or production figures, whichever is the limiting factor and analysed in terms of product. From this is prepared sales budgets, production budgets, expense budgets, labour budgets, and material budgets – and not least but so often forgotten, a finance budget.

It is important that each individual budget be worked out in collaboration with the person responsible for controlling costs and performance in that budget area. Unless there is full co-operation based on confidence and understanding the full value of budgets as an incentive and means of control will be largely lost.

The management accountant will be dependent on the technical knowledge and advice given him by the operating staff, but the more he can learn about the practical problems involved, the better he will be able to help and advise in his work, particularly when he is dealing with smaller concerns.

The following list, which is by no means exhaustive, gives some indication of the matters which require study:

(a) the establishment of sales budgets; this generally requires a rudimentary knowledge, at least, of market research;
(b) the setting of labour standards, which involves a knowledge of how much a man should be able to do in a given period of time at a given type of job; this means being familiar with the techniques of work measurement;
(c) wage rate systems, including incentive bonus schemes;
(d) the establishment of material standards, how materials are used and processed and what wastage should be allowed; in some industries he will have to know about yields from materials and processes; his sources of information will be the technical, production and buying departments;
(e) the behaviour of overhead expenses at varying levels of production;

not initiate, he may see opportunities, but may not seize them, he may detect problems but does not solve them.

He is part of management, but he is not the manager. The difference between line management and staff advisory function must always be observed.

Even within these confines the management accountant's role is still a positive one.

By helping in planning, by recognizing potential, by the analysis and recognition of problems and by sound, factual reporting (not the easiest of tasks) he can materially help management in making sound and workable decisions.

The setting of budgets and standards

The master budget of a business is a grouping of a number of budgets of activities which are co-ordinated within a master plan.

It follows that there must be a business organizational plan and clearly demarked responsibilities of persons for every part of the business.

Too often when management accounts are about to be installed no organizational charts exist and responsibilities of executive staff are poorly delineated and identified.

This need for prior organization will almost certainly throw a strain upon the installing accountant and stresses the need for all accountants to have a good knowledge of business structure and organization.

The first stage in planning is to evaluate the business so as to determine the amount of profit required. This minimum profit has been defined as 'a profit target sufficient to provide normal reserves, replace capital assets, cover fixed interest and permit the payment of an appropriate dividend' (*Management Accounting* – Association of Certified Accountants).

This profit target should therefore be the basis and not merely the result of forward planning, and should be related to the capital employed and the risk element involved in the particular business.

It is important that the capital employed should be thought of in terms of current values not historical expenditure (*The Planning and Measurement of Profit* – Association of Certified Accountants publication.

If, in the preparation of the profit plan, it becomes obvious

CHAPTER VI

THE ACCOUNTANT'S ROLE IN THE EFFECTIVE USE OF MANAGEMENT ACCOUNTS

In undertaking the role of a management accountant the accountant will find himself taking up additional responsibilities and a much greater involvement in the business operation.

It is at this point that the difference in the training and experience of the accountant in industry and commerce and the practising side becomes so very apparent. This difference has been referred to in an earlier chapter.

The practising accountant as a rule is more at home in the boardroom and is usually familiar with the type of information upon which directors and general management make many of their decisions, usually of a financial nature.

The moment decisions are to be made on operational details and when lower levels of management are involved the practising accountant will feel less at home.

The only solution to this difficulty is to become involved with the key personnel and to discuss with them their problems and their needs. This co-operation will not only aid in the earlier preparatory work but will be of material assistance in the effective use of the accounting information subsequently.

Where the practising accountant operates as a consultant and installs the system on a 'once-and-for-all' basis, it is desirable that the audit be extended to cover the review of its working to ensure that it continues to serve the needs of the business and keep up with the new developments.

When the practising accountant undertakes to operate the system, it is desirable that the service be 'personalized' by making it one person's responsibility so that the relationships between management personnel and the management accountant can be developed to mutual understanding and esteem.

The position of the management accountant
The position of the management accountant is much more delicate than that of the financial accountant.

A good management accountant is full of initiative but does

to a more complex form of presentation; perhaps not so easily read or understood.

Many forms of presentation can be developed depending upon ingenuity and the circumstances – however, the rules of comparison and clear presentation must always be observed.

Terminology

The choice of terminology can often present a problem for the accountant.

Business language undoubtedly varies by locality, industry and even to the degree of modernity of the business concerned.

The accountant must obviously address himself to the recipients of his returns, in language which is understood.

Jargon, of course, is unforgivable. Simple clear phraseology is always desirable.

But where accounting terms are concerned it may be preferable to try and obtain the use of approved forms.

The Institute of Cost and Management Accountants publish through Gee & Co, *Terminology of Cost Accounting*. This can be useful as a guide although the conversion to an approved terminology may require patience, dual titles and a training of the recipients.

THE PRESENTATION OF INFORMATION 37

Exhibit No. 15

RECONCILIATION OF ACTUAL PROFIT WITH BUDGETED PROFIT

	March £	March £	Year to Date £	Year to Date £
1. Our BUDGETED PROFIT was		49,173		368,316
2. Our ACTUAL TRADING PROFIT was (Exhibit 14, line 8)		52,927		367,393
3. Therefore our ACTUAL PROFIT was higher/*lower** by		£3,754		£923*
The VOLUME and MIXTURE of SALES was more profitable than the BUDGET and produced:				
4. (a) An additional STANDARD PROFIT of but (b) SALES DISCOUNTS were *higher** by and (c) ALLOWANCES had to be given of ..		4,070 4,206* 307*		9,971 17,903* 1,826*
5. The net effect being		443*		9,758*
(d) We also recovered more/*less** than we anticipated towards ADVERTISING by and SELLING and ADMINISTRATION EXPENSES by	597 1,315		22* 2,206	
		1,912		2,184
6. The VOLUME of FACTORY PRODUCTION was lower than expected, resulting in an *increase** in production costs of ..		175*		4,636*
7. The total variation from BUDGET due to VOLUME was		1,294		12,210*
The following items show how the ACTUAL COSTS of MANUFACTURING, SELLING and ADMINISTRATION compared with the BUDGETED COSTS:				
(a) FACTORIES: LABOUR COSTS were *higher** by .. MATERIALS USED were lower by OVERHEAD COSTS were lower by	192* 764 414		2,081* 3,640 3,262	
		986		4,821
(b) SELLING and ADMINISTRATION: BRANCH COSTS were *higher** by .. HEAD OFFICE COSTS were *higher** by TRAVELLERS' COMMISSIONS not payable were	355* 220* 834		2,971* 424* 3,333	
		259		62*
(c) PURCHASED MATERIALS and GOODS cost less by		1,215		6,528
8. The total of all these items accounts for the rise/*fall** in our ACTUAL TRADING PROFIT of (line 3 above)		£3,754		£923*

Exhibit No. 14
MONTHLY PROFIT AND LOSS ACCOUNT

	March		Year to Date	
	£	£	£	£
1. The ACTUAL NET SALES, less discounts, bonuses, rebates and allowances were		269,153		2,101,888
2. From these sales we deduct the STANDARD COSTS for: MANUFACTURED and PURCHASED GOODS, ADVERTISING, TRAVELLERS' COMMISSION and TRANSPORT		201,730		1,606,220
3. This leaves a GROSS MARGIN of		67,423		495,668
4. We then deduct the recovered STANDARD COSTS for SELLING and ADMINISTRATION EXPENSES		18,693		137,110
5. Our STANDARD PROFIT from ACTUAL SALES is therefore		48,730		358,558
6. The prices paid for RAW MATERIALS and PURCHASED GOODS were, however, less than budgeted, resulting in an additional profit of		1,215		6,528
7. This profit is also increased/*decreased** by the difference between the ACTUAL COSTS incurred and the STANDARD COSTS recovered from SALES (2 and 4 above) of the:				
FACTORIES	811		185	
ADVERTISING	597		22*	
TRAVELLERS' COMMISSION	834		3,333	
SELLING and ADMINISTRATION	740		1,189*	
		2,982		2,307
8. The ACTUAL PROFIT on TRADING is		52,927		367,393
Other revenue/*expenses** which affect the PROFIT are:				
9. EXPENDITURE on RESEARCH and DEVELOPMENT	1,224*		9,572*	
10. MISCELLANEOUS REVENUE	363		1,570	
11. MISCELLANEOUS EXPENDITURE	457*		4,848*	
A net *decrease** in profit of		1,318*		12,850*
12. The net result of ALL the COMPANY'S TRADING ACTIVITIES is a NET PROFIT before taxation of		£51,609		£354,543

LOSS ACCOUNT – Period 10   *Exhibit No 13*

COMPARISON WITH STANDARD

			£	£	£
STANDARD NET SALES				301,698	
Selling prices variance	Gain	+		5,789	
Volume variance	Gain	+		1,500	
Bonus and rebates variance	Loss	−		3,295	
NET SALES					305,692
FACTORY STANDARD COST OF PRODUCTION				198,760	
Adjust for factory variances:					
Material prices	Loss	+	1,074		
Material usage	Loss	+	471		
Wage rates	Loss	+	290		
Labour efficiency	Loss	+	370		
Overheads – Volume	Gain	−	326		
Spending	Loss	+	524		
	Loss	+		2,403	
ACTUAL PRODUCTION COST				201,163	
STANDARD FREIGHT TO BRANCHES			3,064		
Freight to branches variance	Loss	+	293		
				3,357	
PRODUCTION AT COST TO BRANCHES				204,520	
Less Increase of STOCK				25,210	
					179,310
STANDARD COST OF GOODS PURCHASED			42,900		
Adjust for price variance	Loss	+	478		
				43,378	
Less Increase of STOCK				2,375	
					41,003
STANDARD COST OF SELLING AND BRANCH OPERATING				39,509	
Adjust for variances – Volume	Gain	−	1,009		
Spending	Gain	−	2,870		
				3,879	
					35,630
STANDARD COST OF ADMINISTRATION				13,507	
Adjust for variances – Volume	Gain	−	385		
Spending	Gain	−	493		
				878	
					12,629
					268,572
ACTUAL TRADING PROFIT					£37,120

MONTHLY PROFIT AND

PROFIT AND LOSS ACCOUNT

	£	£	£
SALES			317,916
Less Bonus and rebates			12,224
NET SALES			305,692
Less			
FACTORY COST OF GOODS SOLD (including freight to branches)		179,310	
COST OF PURCHASED GOODS SOLD		41,003	
COST OF SELLING AND BRANCH OPERATING		35,630	
COST OF ADMINISTRATION		12,629	
TOTAL COST OF GOODS SOLD			268,572
ACTUAL TRADING PROFIT			37,120
Add Miscellaneous income			1,238
TOTAL INCOME			38,358
Less Miscellaneous expenditure			1,071
NET PROFIT			£37,287

It is also very important at this level to differentiate between labour costs and material costs classifying them between controllable and uncontrollable (by supervisor) otherwise action upon figures cannot easily be taken.

Management by objectives

Mr John W. Humble in his booklet *Management by Objectives* (Industrial Educational and Research Foundation) states that: 'it is very necessary for a company to relate the overall objectives and responsibilities of the organization to that of the individual manager'.

Management by objectives (M.B.O.) is a corollary to the principle of accounting for responsibility, one of the basic precepts of management accounting.

It therefore behoves the alert and managerial minded accountant to play an active part in the conduct of a system of M.B.O. By his position and work and the management accounting information available to him he can make a substantial contribution to management's effort in control through this technique.

Examples of management returns

Exhibit No. 13 shows on the left-hand side a conventional profit and loss account, and on the right, the profit and loss account in standard cost form. It will be noted that the figures shown in the conventional profit and loss account correspond at various stages with those shown on the standard cost section of the account. This dual form of presentation is useful when some of the board of directors are more used to the conventional forms of accounts and as a bridging procedure upon the introduction of standard cost forms of presentation.

The second example is that of profit and loss account in standard cost form marked Exhibit No. 14, and second sheet marked Exhibit No. 15, giving a reconciliation between the actual and the budgeted profit for the period. One of the virtues of this example is that the explanatory information is set out in narrative form.

Exhibits No. 14 and 15 could, of course, be combined as it would have been possible to show both actual and target figures in the profit and loss account itself. This would have led

Continued on page 38

Reports to top management
As it is the responsibility of top management to run the business it is appropriate that the information given should cover all phases of the organization.

It should be remembered that management involves delegation of authority and responsibility.

The returns to top management therefore should really consist of a summary of statistics and reports to departmental managers. Top management is therefore able to spot the strength and weakness of the parts of the organization and take action through the departmental heads concerned.

Reports to departmental management
It is at this point that the principle of accounting for responsibility is clearly recognizable.

Each department head is responsible for and must account for the success or failure in the operation of his department.

Departmental accounts may consist of sales, turnover, or expense budgets upon which the departmental head must stand or fall.

Departmental accounts therefore serve three functions. Firstly, to assist the departmental head in running his department, secondly, to permit top management to develop its functions to observe and guide the departmental head and thirdly, to facilitate top management in co-ordinating the separate departments and aspects of the business.

Reports to shop management
With the greater application in business of the concepts of the behavioural scientist, the need for giving the supervisors information and obtaining their commitments to targets of performance has emerged. There is now little disagreement that the supervisor is a manager although the circumstances under which he works is often not conducive to the full use of his potential.

At this level of management where the supervisor is closer to the material activities of the business, it is often found that quantity statistics other than financial are of great importance.

Man-hours, absentee statistics, scrap percentages are often more meaningful in plain figures than converted to pounds sterling.

It may appear that these rules set too high a standard. In practice, this cannot be so. In too many cases, the information submitted creates a sense of distaste, irritation and even frustration.

The aim should be to make the acquisition of control information a pleasant and absorbing experience with a sense of mastery upon completion.

Even more important than the form of presentation is timing and selection of information for presentation. A company chairman may be prepared to examine the figures available over each four-week period to satisfy himself that the policies determined upon are being carried out. To do so more frequently would be no advantage and may be confusing.

The foreman requires to know the results weekly and within a day or so, and in certain cases production methods make daily information absolutely necessary.

It is a common complaint since the introduction of computers that the recipients of control information are smothered by a mass of information, which it is frankly admitted is not read. The fact that a computer can spew out large quantities of facts and statistics merely throws upon the accountant greater responsibility for not giving unnecessary information.

This skill of the accountant must be exercised to present only the information which is necessary for the recipient's task, to present it intelligently and effectively.

Information is valueless without this skill of the accountant in communicating it when and how it is needed.

The reports and statements presented in a business by an accountant usually fall into four classes – these are: boardroom figures, top management reports, middle or departmental management returns, and information for supervisors.

Boardroom figures

The reports and statements for presentation to the directors as members of the board should be global in coverage, without details required for running the business, but to show clearly the sources and level of profits earned, the financial situation and capital expenditure. It is important to draw to the board's attention any trends which show a variation from the normal or planned course of business and so effect policy making.

Chapter V

THE PRESENTATION OF INFORMATION

The presentation of information to management and supervision is a distinct and different problem to that of operating an accounting system or the compilation of information.

The cooking of a meal is not the same as serving it attractively on a plate on a well laid out table. The presentation of figures also requires to be in a form to satisfy the recipient's needs and to stimulate his appetite.

Layout is important. Who understands this better than the advertising layout man? His consciousness of quality, style, appeal and clarity are worthy of study by the accountant.

The layout should be appropriate to the conscious and even unconscious needs of the reader. Few people enjoy figures for their own sake and if the manager did it might well be he was not a good manager! So each according to his need – the needs of the chairman are very different to the requirements of the supervisor.

The more obvious rules to apply are as follows:

(*a*) Attractiveness in appearance.

(*b*) Quality and dignity appropriate to the occasion and reader.

(*c*) Good layout, aiding both eye and mind to absorb the information.

(*d*) Information applicable to the responsibilities of and in terms understood by the recipient.

(*e*) Constant form so that a habit of perusal and thought is built up.

(*f*) As far as possible one aspect of information to one sheet of paper – it aids instant recollection.

(*g*) No figure given without supplying some comparison to give it value; targets are the best form of comparison whether in graph form or figures.

(*h*) Form design prepared in conjunction with recipient or class of recipient.

(*i*) Up-to-date information, and

(*j*) Reliability and accuracy.

MANAGEMENT STATISTICS

Exhibit No. 12

	This month	Last month	Three months to March 31st, 1972
1. Total distribution cost as percentage of sales	21·6	20·8	20·9
2. Net profit before taxation as percentage of sales	5·1	5·4	6·1
3. Net profit before taxation as percentage of total net assets	1·1	1·3	4·1
4. Ratio of net current assets to current liabilities	1·06	1·02	—
5. Ratio of sales to total assets	2·8	2·8	—
6. Debtors – number of days' turnover:			
Home	55	54	—
Export	48	49	—
All trade debtors	53	52	—
7. Stocks of raw materials:			
Number of days' consumption	57	54	—
Stocks of finished goods:			
Number of days sales –			
Manufactured	20	18	—
Merchandise	61	65	—
8. Order book position:			
Manufactured goods	£173,400	£160,900	—
Merchandise	£31,300	£25,200	—
Number of days' production represented by orders for manufactured goods	78	73	—
9. Yield on total assets:			
(a) Manufacture:			
Yield as percentage of capital employed	2·9	—	8·9
Which is at the annual rate of	34·8	—	35·6
(b) Distribution:			
Yield as percentage of capital employed	·16	—	·33
Which is at the annual rate of	1·92	—	1·32
(c) Manufacture and distribution:			
Yield as percentage of capital employed	·79	—	2·91
Which is at the annual rate of	9·48	—	11·64

Exhibit No. 11

STOCK BUDGET
AUGUST–SEPTEMBER 1972

Description	Unit	Estimated consumption Aug./Sept. 1972	Plus budgeted closing stock	Total	Less opening stock	Net budget purchases for Aug./Sept. 1972	
						Quantity	Value
							£
Oil	gal.	230	10	240	14	226	254
Jam	cwt	50	2	52	2	50	211
Sugar:							
Granulated ..	cwt	3	1	4	1½	2½	9
Castor	cwt	150	10	160	1	159	560
Icing	cwt	12	5	17	1	16	69
Margarine ..	cwt	200	10	210	6	204	1,503
Butter	lb.	500	60	560	27	533	93
Cream:							
'A' Cream ..	lb.	4,200	100	4,300	304	3,996	333
'B' Cream ..	gal.	1,190	60	1,250	36	1,214	668
Savoury mix ..	lb.	2,200	200	2,400	410	1,990	125
Sausage meat ..	lb.	3,700	200	3,900	12	3,888	356
Flour	bags	240	20	260	8	252	655
'D' Flour	lb.	8,600	200	8,800	110	8,690	435
Eggs:							
Liquid	lb.	5,850	450	6,300	308	5,992	549
Dried	lb.	800	—	800	—	800	270
Bakers' chocolate	lb.	2,500	50	2,550	1,061	1,489	124
Coconut	lb.	650	—	650	60	590	26
Currants, sultanas	lb.	4,040	560	4,600	1,130	3,470	174
Fondant	cwt	80	—	80	5	75	288
Piping jelly ..	cwt	36	3	39	2	37	233
Marzipan	cwt	31	3	34	10	24	397
Nuts	lb.	380	55	435	131	304	29
Ground almonds ..	lb.	460	—	460	142	318	80
Meringue mixture	cwt	7	½	7½	½	7	75
Gateaux plates ..	gross	44	—	44	—	44	18
Baking cases ..	thous.	272	16	288	40	248	43
							£7,577

There are many types of ratio which may be employed, some better known or more obviously useful than others. The choice of ratios and statistics to be presented should be the result of a careful study of the operating patterns of a business and the figures which will give the control required.

The examination of trends in ratios, and moving annual totals over a period of years may expose important trends making for greater or less success in business operation.

figures can be seen undistorted by minor short-term factors.

Cash forecasts and movement of funds statements seem to be little understood or valued in many places.

In these days of substantial inflation there is always a serious lack of cash if the organization is maintained at a level or increasing rate of activity.

Replacement cost accounting, which was once a dirty word is now under serious reconsideration by authoritative accounting sources. Recent examples of large companies, of great reputation, failing through lack of funds, must leave all with the impression that there was insufficient planning and provision in respect of financial resources.

There are, indeed, prominent members of the accounting profession who advocate depending less on the orthodox profit and loss account, and planning and controlling a business on cash flow statements.

The importance of controlling the cash position, present and future, cannot be questioned.

Stock budget (Exhibit No. 11)

This is an example of a simple stock budget applied to a bakery in this instance, the principles of which may be applied to similar needs in a variety of industries.

Such a budget enables management to obtain departmental control of stocks, to compare actual with target stocks, and to take action on major discrepancies.

This form of stock budget prevents the excessive growth of stock items either by accident or by lack of control and guides the buyers in making their purchases for materials to be delivered during the period covered by the stock budget.

If a business requires a comparatively large or complex holding of stock, the situation may require more than a stock budget.

Streamlining and simplification of product, simplification and standardization of components and raw materials may make substantial cost savings in investment and in control of stock.

Management statistics (Exhibit No. 12)

This exhibit is an example of the type of control statistics which can be prepared to help management to become aware of the important facts and trends of the business.

Exhibit No. 10

MOVEMENT OF FUNDS STATEMENT

	6 months ended June 30th, 1972		Year ended June 30th, 1972		5 years ended June 30th, 1972	
	£	£	£	£	£	£
FUNDS AVAILABLE:						
Profit before taxation per profit and loss account		46,285		73,615		245,206
Depreciation charged in accounts		4,920		8,706		35,132
Increase/*Decrease* in bank overdraft		18,341*		11,236*		88,041
Increase/*Decrease* in borrowings from subsidiary companies		26,429		11,623*		8,514
Increase/*Decrease* in trade creditors		25,284*		19,194*		44,107
Sales of fixed assets		—		—		3,100
Increase in share capital		—		50,000		50,000
TOTAL FUNDS AVAILABLE FROM ALL SOURCES		34,009		90,268		474,100
DEDUCT AMOUNTS APPLIED IN:						
Increase in stocks	6,623		12,130		87,798	
Increase/*Decrease* in trade debtors	17,928*		11,628		62,803	
Increased investment in subsidiary company	—		—		41,250	
Payments of taxes	25,398		25,398		104,373	
		14,093		49,156		296,224
BALANCE, BEING EXPENDITURE ON FIXED ASSETS		£19,916		£41,112		£177,876

finished stock, and the payment by debtors, as these are effected by variables outside of absolute control.

However, strict stock control arrangements and a good follow-up of debtors can usually bring the situation under reasonable control.

In Exhibit No. 9, the first forecast is made in ROMAN numerals. As the first forecast showed a maximum cash deficiency of just over £117,000 a readjustment of the plan was made to reduce the deficiency to £110,000 which is the limit prescribed by the company's bankers.

The amended forecast is shown in ITALIC figures.

This process might be described as cutting one's coat to fit one's cloth.

Like all plans, it serves no useful purpose unless used as a guide for subsequent performance.

The cash forecast statement (Exhibit No. 9) can be developed into a monthly and accumulated report on the cash position showing the variance against the plan.

The monthly cash statement would compare the cash forecast figures against the actual figures arising, showing the plus and minus variations. Thus any cash movements away from the plan can be speedily observed and appropriate action taken to control any undesirable trends that may be developing.

Movement of funds statement (Exhibit No. 10)
This is a statement which most businesses need but surprisingly few seem to prepare.

There are a number of ways of drawing up a movement of funds statement, but the one which has been included in this book starts off by arriving at the total of the funds available for disposal and then goes on to show how they have been utilized. A feature of this statement is that it ends up with the item in which the company is particularly interested for the time being – in this case the expenditure on fixed assets. If the problem had been to build up stocks out of available resources, then the statement would have ended with the increase in the amount invested in stocks. It should, perhaps, be pointed out that no dividends have been paid by the company during the period covered by this statement.

It is important that a movement of funds statement should cover a sufficient period of time to ensure that the trend of the

CASH FORECAST

Twelve months January to December 1972

DATA	Jan.	Feb.	March	April	May	June	J
Overdraft limit £110,000	£	£	£	£	£	£	
Sales during month	40,000	30,000	30,000	30,000	45,000	50,000	60
						50,000	*50,*
Debtors at end of month	80,000	70,000	60,000	60,000	75,000	95,000	110,
						95,000	*100,*
Receipts from debtors	40,000	40,000	40,000	30,000	30,000	30,000	45,
						30,000	*45,(*
Raw materials consumed	20,000	15,000	15,000	15,000	22,500	25,000	30,(
						25,000	*25,(*
Minimum stocks of raw materials	30,000	30,000	30,000	40,000	40,000	45,000	45,0
						40,000	*45,0*
Purchases of raw materials	20,000	15,000	15,000	25,000	22,500	30,000	30,0
						25,000	*30,0(*
FORECAST							
Payments:							
Creditors for raw materials	20,000	20,000	15,000	15,000	25,000	22,500	30,00
						22,500	*25,00(*
Wages	10,000	10,000	10,000	10,000	11,250	12,500	15,00(
						12,500	*12,50(*
Creditors for general expenses	5,000	5,000	5,000	5,000	5,000	5,000	5,00(
						5,000	*5,000*
Taxes	30,000	—	—	—	—	—	—
Creditors for capital expenditure	—	10,000	10,000	10,000	20,000	—	—
Total cash expenditure	65,000	45,000	40,000	40,000	61,250	40,000	50,000
						40,000	*42,500*
Receipts from debtors	40,000	40,000	40,000	30,000	30,000	30,000	45,000
						30,000	*45,000*
Cash deficit for period	25,000	5,000	—	10,000	31,250	10,000	5,000
						10,000	—
Cash surplus for period	—	—	—	—	—	—	—
						—	*2,500*
Overdraft brought forward	25,000	50,000	55,000	55,000	65,000	96,250	106,250
						96,250	*106,250*
Total overdraft requirement at end of month	50,000	55,000	55,000	65,000	96,250	106,250	111,250
						106,250	*103,750*

(NOTE. – Italic figures indicate forecast revised in order to keep the overdraft within the lim

Chapter IV

MANAGEMENT ACCOUNTING STATEMENTS – CONTINUED

The previous chapter was given over to the control of expense, output and profit by management accounting techniques.

This chapter treats with the application of accounting control to money availability and use.

It is in the area of money control that management is often most tested. Many a business with considerable potential has tumbled because the financial needs within the company were not related to the money available without. Careful control of financial resources is the keystone to successful development of an organization

Cash forecast and control (Exhibit No. 9) is an orthodox cash forecast. The upper part of this forecast headed 'Data' sets out some of the basic information from which the forecast is prepared.

An accounting statement cannot be better than the information on which it is based, and as a cash forecast requires the use of figures arising from a forecast activity month by month for a considerable forward period of time, very great care in preparation, knowledge of events and awareness of current trends is necessary.

The primary requirements of a cash forecast are as follows:

(a) A reasonable estimate of profits, broken down into elements of expense and revenue. When a full system of forecasting and budgetary control is operating this information should be readily available.

(b) A capital expenditure budget. Although cash forecasting usually needs only a fairly short-term capital budget, the actual plan of machinery, plant and building acquisition and replacement should cover several years ahead.

(c) An estimate of tax payable on the forecast profit.

It is usually fairly easy to plan the timing and amount of expenditure as this is within control. It is much more difficult to estimate the holdings of raw materials, components and

The standard recovery of selling and administration is the product of the selling and administration recovery rate applied to the works standard cost of sales.

The volume variance in column 5 is the over- or under-recovery of all expenses due to an over- or under-volume of work (in the tool-room and maintenance departments) or of sales compared with the budget. The volume variance on this exhibit is not confined to the over- or under-recovery of fixed expenses as it is for the production departments.

The spending variance in column 6 is the difference between the allowed expenses (in this case the budgeted expenses) and the actual expenses of the departments; the actual expenses are determined in the same way as for the production departments.

Exhibit No. 8 completes the set of monthly accounts which may be said to deal with the profitability and efficiency of the concern.

Other important reports and statements required by management for complete control are dealt with in Chapter IV. They include statements of capital expenditure, cash position and the order book, etc.

Exhibits Nos 1 to 8 are tailored to meet a typical company's requirements bearing in mind the type of business, its size and management requirements.

Whatever the type of business, the principles remain the same, but the application may vary according to needs of the concern.

The basic principle is to supply the type of information required to guide the management to co-ordinate the business, to provide a means to measuring the progress of the business. This will involve an examination of the organization, a determination of the responsibilities of departmental heads, etc., a recognition of the technical problems and special trade problems, such as seasonal variations of production and sales.

As management philosophies and applications change and developments occur in organization and operation of a business so will the information needs of the business change. Management accountants must at all times be ready to meet the changing situation by making suitable adjustments in the form of information prepared and supplied.

Exhibit No. 8

SUMMARY OF SERVICE, SELLING AND ADMINISTRATION DEPARTMENTS' EXPENSES

	1	2	3	4 (2–3)	5 (2–1)	6 (1–3)
	Allowed expenses (as budget)	Standard recovery	Actual expenses	Total variance Gain/Loss*	Analysis of variance Volume Gain/Loss*	Spending Gain/Loss*
	£	£	£	£	£	£
Five weeks to June 30th, 1972 Period 6						
SERVICE DEPARTMENTS						
Tool-room	7,036	5,160	6,172	1,012*	1,876*	864
Maintenance	1,368	1,190	1,186	4	178*	182
Stores and dispatch	1,195	1,195	1,402	207*	—	207*
Works general	3,199	3,199	3,408	209*	—	209*
	£12,798	£10,744	£12,168	£1,424*	£2,054*	£630
SELLING AND ADMINISTRATION	£6,706	£6,310	£8,220	£1,910*	£396*	£1,514*
Twenty-six weeks to June 30th, 1972 Periods 1–6						
SERVICE DEPARTMENTS						
Tool-room	36,589	28,740	30,790	2,050*	7,849*	5,799
Maintenance	7,115	6,226	6,112	114	889*	1,003
Stores and dispatch	6,216	6,216	7,188	972*	—	972*
Works general	16,635	16,635	15,572	1,063	—	1,063
	£66,555	£57,817	£59,662	£1,845*	£8,738*	£6,893
SELLING AND ADMINISTRATION	£34,872	£31,076	£5,623	£4,547*	£3,796*	£751*

MANAGEMENT ACCOUNTING STATEMENTS FOR A SMALL CONCERN 21

is also included on the statement as there is a charge for inspection in the standard works cost of all products.

The statement is designed primarily for use by the works manager, and it is built up by summarizing the detailed operating statements (Exhibit No. 5) of the various works departments.

The total variances shown in the last column of the statement are those which appear in the profit and loss account (Exhibit No. 1).

Summary of service, selling, and administrative departments' expenses (*Exhibit No. 8*)

This statement shows the total variances arising on the expenditure of the service, selling, and administrative departments; it also analyses the total variances between volume and spending variances.

No adjustment has been made to the budgeted expenses of each department when computing the figures of the allowed expenses for the period in column 1. The concept of 'levels of activity', which was used in connection with the operating statements for the production departments, does not strictly apply. The work of the tool-room and maintenance departments can be measured in terms of wages incurred in serving other departments and, in fact, is so measured when charging out to these other departments; but in determining the proper level of expenses in the tool-room and maintenance departments the amount of wages paid is not a reliable guide. The expenditure of the stores and dispatch and works general is almost entirely of a fixed nature, and no convenient yardstick of activity exists. Similarly, selling and administration expenditure is largely of a fixed nature with a few big exceptions. The exceptions are discounts and commissions payable and carriage outwards, but it was not considered practicable at this stage to make adjustments only for these items when the figures for all the other service departments were according to the budget.

The standard recovery in column 2 represents, in the case of the service departments, the charges made to other departments. As the stores and dispatch and the works general departments charge out sums equal to their budget, the figures in columns 1 and 2 are the same and so no volume variance arises in column 5, having been dealt with in full in the operating accounts of the production departments.

Five weeks to J
Period 6
1. ACTIV
2. LABOU

   WAGE
3. Sta
4. Act
5. Tot
   Var
6. R
7. E

   RAW
8. Sta
9. Act
10. Var

   EXPEN
11. All
12. Sta
13. Act
14. Tot
   Var
15. V
16. S

TOTAL STA

Twenty-six wee
Periods 1–6
1. ACTIV
2. LABOU

   WAGE
3. Sta
4. Act
5. Tot
   Var
6. R
7. E

   RAW
8. Sta
9. Act
10. Var

   EXPEN
11. All
12. Sta
13. Act
14. Tot
   Var
15. V
16. S

TOTAL STA

Exhibit No. 7
facing page 20

The fourth column is headed 'actual expenses'. In the case of direct charges to the departments the amounts are either allocations or apportionments of the actual expenses incurred in the period as shown in the various expense accounts in the general ledger. For example, the actual costs of waiting time and indirect labour are based on an analysis of the wages sheets for the period; the 'actual' cost of gas is the estimated proportion for this department of the total gas consumption of the factory during the period. The charges for the service departments are determined differently. The maintenance department and tool-room charge all departments for whom they work the actual wages incurred on the various jobs in the period, plus a departmental recovery rate to recover their own departmental expenses. The works general and stores and dispatch departments each make a fixed weekly charge and this, as it is calculated on the overhead expenses budget, is the same as the budgeted expense in column 1. Since the charge is treated as fixed it is also the same as the allowed expense in column 2.

The other variance shown on this statement in column 7 is the difference between 'allowed expenses' and actual expenses. As has already been explained, allowed expenses incorporate an adjustment for the level of activity. Thus, when they are compared with the actual expenses for the period, the difference represents over-spending or under-spending; for this reason the variance is referred to, somewhat aptly, as a 'spending variance'.

This spending variance does not arise on the charges from the works general and stores and dispatch departments. By making fixed charges to the production departments based on budget, the spending variances are left in these service departments' own operating accounts and are shown in full in the summary of service, selling and administration departments – Exhibit No. 8.

In order to reduce the number of figures which a foreman has to study each month, the details of cumulative results have been omitted, and only the cumulative totals are shown on the schedule of expenses.

Summary of works operating departments (Exhibit No. 7)
This statement provides an overall picture of the operations of all the production departments. The inspection department

MANAGEMENT ACCOUNTING STATEMENTS FOR A SMALL CONCERN 19

this is a practical approach rather than one which is theoretically correct. This is also true of the division of expenses between those which are considered to be within the control of the foremen and those which are not.

It was not practical at the beginning of the scheme to establish budgets for the semi-variable expenses for various levels of production; all expenses, therefore, had to be considered as either fixed, directly variable or, in a few instances, partly fixed and partly directly variable in assumed proportions, e.g., one-half fixed, one-half directly variable.

Column 2 shows the 'allowed expenses' which are the permitted expenses at the actual level of activity for period 6. As this is 80 per cent of the budgeted production, the allowed expenses equal 80 per cent of the budgeted variable expenses and the variable portion of the semi-variable expenses, plus the whole of the budgeted fixed expenses and the fixed portion of the semi-variable expenses.

The next column in the statement is headed 'standard recovery'. This is the amount of overhead expenses included in and thus 'recovered' in the standard cost of actual production for the period. The method used to calculate the overhead cost recovered is described in Chapter VII in the section dealing with the overhead expenses budget. There are various methods but they are all based on the principle that the budgeted overheads of a department are fully recovered in the standard cost of production, provided production is at the budgeted level. If actual production exceeds budget the overhead cost will be more than fully recovered; but if actual production is less than budget, say only 80 per cent, then only 80 per cent of the budgeted fixed and variable overheads of a department will be recovered. It will therefore be seen that the difference between 'allowed expenses' and 'standard recovery' is the volume variance, which is shown in column 6; this variance represents the over- or under-recovery of fixed expenses in the cost of production resulting from a difference between actual and budgeted production. In other words, if production is 80 per cent of the budget, the allowed expenses will consist of 100 per cent of the budgeted fixed expenses and 80 per cent of the directly variable expenses. The standard recovery will amount to 80 per cent of both fixed and variable expenses. Therefore 20 per cent of the fixed expenses will not be recovered.

solvent extracts, but one advantage is that it is sufficiently sensitive to be applied directly to water samples, allowing measurement of what has become known as the 'in vivo' fluorescence. This direct measurement may be made in the laboratory or in the field, since instruments are now available with flow-through cells which allow water to be pumped directly to the instrument, greatly facilitating direct field measurements, particularly during investigations of horizontal variability in the phytoplankton.

The method was developed by Lorenzen (1966) and the assumptions made and the errors therein have been investigated more recently by Kieffer (1973a, b), Loftus & Seliger (1975) and Heaney (1978). The *in vivo* fluorescence is calibrated by comparison with the absorbance of a solvent extract of a given sample. The method assumes a constant ratio between the emission intensity of the *in vivo* sample and the extractable chlorophyll *a*, yet the range of variation in this ratio has been shown to be tenfold in natural phytoplankton samples. This variation may have several causes, including differences in emission intensity between planktonic species, interference or quenching by solutes, light scattering by particles, changes in fluorescence with temperature, and ambient light fluorescence inhibition. Some of these problems may be overcome, at least in part, by investigation of the emission characteristics of the individual algal species from the waters of interest, using controls of filtered water, measurement at constant temperature and dark adaptation of the plankton samples for a fixed time. Yield fluctuations may also be eliminated by addition of 10 μM DCMU (3–(3,4 dichlorophenyl)–1,1–dimethylurea) to the cell suspension (Slovacek & Hannan 1977). Then, it is claimed, the *in vivo* fluorescence yield becomes maximal and a constant function of cellular chlorophyll *a*, regardless of growth conditions or of the species examined.

Any spectrofluorimeter fitted with a photomultiplier tube suitable for detection of the longer wavelengths emitted by chlorophyll is suitable for laboratory measurement of chlorophyll *a*, and the other chlorophylls (Loftus & Carpenter 1971). For field as well as laboratory measurements, the most popular instrument appears to be the Turner III flow-cell Fluorometer. The version used successfully at this laboratory is equipped with a blue lamp (110-853) and the red-sensitive photomultiplier tube R136. A Corning CS 5-60 filter is used for excitation, providing maximum excitation at 425 nm. Emitted fluorescence is measured through a sharp cut-off filter (Corning, CS 2–64) with maximum transmission at 687 nm.

If an instrument with a grating monochrometer or interference filters is available, it is possible to distinguish the accessory pigment fluorescence of different groups, e.g. green algae (λ_{ex} 478 nm, λ_{em} 678 nm) and some blue greens (λ_{ex} 615 nm, λ_{em} 647 nm) in mixed populations (Caldwell 1977).

The measurement of *in vivo* chlorophyll fluorescence is enjoying increasing popularity as a method for studying planktonic populations, but problems caused by light scattering and interference will probably limit the fluorescence investigation of benthic material to solvent extracts (a method devised for soil is given by Sharabi & Pramer 1973).

6.4.1.2 *Bacteriochlorophylls*

Bacteriochlorophylls may be detected in the presence of algal chlorophyll *a* by the nature of their distinctive absorption spectra. The purple bacteria possess bacteriochlorophylls *a* and *b* (absorption maxima 850 nm and 1000 nm), the green bacteria possess bacteriochlorophylls *c* and *d* with absorption maxima at 700–750 nm (previously called *Chlorobium* chlorophylls 650 and 660), plus traces of bacteriochlorophyll *a*.

a. Acetone extraction causes a shift in the absorption maxima of bacteriochlorophylls, usually by about 100 nm, towards shorter wavelengths. Takahashi & Ichimura (1968) and Hussainy (1972) filtered water through cellulose ester membranes which were then extracted with 90% acetone. Ultrasonication for 10 min aids extraction. The concentration of bacteriochlorophyll *a* and *b* [BChl*ab*] is given by the formula

$$[\text{BChl}ab] \ (\mu g \ l^{-1}) = \frac{V_e}{V_s} \cdot \frac{25 \cdot 2}{\ell} \cdot A^{850}_{772}$$

where A^{850}_{772} is the optical density measured at 772 nm corrected by subtracting that at 850 nm.

The concentration of bacteriochlorophylls *c* and *d* are given by the formulae

$$[\text{BChl}c] \ (\mu g \ l^{-1}) = \frac{V_e}{V_s} \cdot \frac{10 \cdot 2}{\ell} \cdot A^{850}_{654}$$

and

$$[\text{BChl}d] \ (\mu g \ l^{-1}) = \frac{V_e}{V_s} \cdot \frac{10 \cdot 8}{\ell} \cdot A^{850}_{662}$$

If sufficiently rich systems are being studied then the plankton may be centrifuged and the pellet used for the determinations. Caldwell & Tiedje (1975b) successfully extracted such pellets with H_2S-saturated acetone and then clarified the extract by recentrifugation or membrane filtration. The absorption spectra of the extracts were scanned between 400 nm and 700 nm and the bacteriochlorophylls determined using the method of Stanier & Smith (1960).

b. *In vivo spectrophotometric determination.* Adapting a method developed by Yentsch for algal chlorophylls, Trüper & Yentsch (1967) were able to examine the *in vivo* absorption spectra of photosynthetic bacteria. The samples are filtered and the filter is mounted in the spectrophotometer lightpath and scanned against a similarly-treated blank filter. For the shorter-wavelength bacteriochlorophylls of the green photosynthetic bacteria, cleared membrane filters may be used. Because these absorb much of the light at wavelengths greater than 750 nm, bacteriochlorophylls *a* and *b* must be determined on glass-fibre filters. The wet filters are mounted in a suitable holder and measured directly against a blank in the range 350-1200 nm. Although there may be problems in obtaining reliable absolute data by this method, it can provide comparative information.

c. *In vivo fluorescence determination.* It is possible to distinguish the fluorescence spectra of the Rhodospirillaceae (λ_{ex} 355 nm, λ_{em} 430 nm) and the Chlorobiaceae (λ_{ex} 470 nm, λ_{em} 773 nm). The relationship between fluorescence intensity and cell numbers is linear over about 4 orders of magnitude and down to 10^4 cells ml^{-1}. However, careful *in situ* controls are needed for each population because the spectra vary with growth conditions (Caldwell 1977).

6.4.2 Microbial cell wall materials

Cell-wall materials do not represent the ideal biomass indicator because they are largely polymeric in nature. Such materials are often slowly decomposed and may survive long after the death of the cell. In spite of this, the analysis of chemicals derived from fungal and bacterial cell walls may be of some limited use and may provide some comparative information where no other methods exist. The algal cell-wall polymer, cellulose, is hardly unique to algae, and therefore serves no useful purpose as a biomass indicator.

6.4.2.1 Fungal biomass

The cell-wall polymer chitin has been used for estimation of fungal biomass in soil and water (Swift 1973a; Willoughby 1978). Although suitable for soil fungi, the method applies in fresh water only to aquatic hyphomycetes; (cellulose is the cell-wall polymer of the Phycomycetes). The method is based on the determination of chitin, either as glucosamine after chemical hydrolysis, or n-acetyl glucosamine after enzymatic cleavage. Chitin is a high-molecular-weight ($c.$ 4×10^5) linear chain of $\beta1:4$ linked N-acetyl-D-glucosamine, and as such its rate of decomposition in the natural environment is likely to be slow. This should be borne in mind when chitin is used as a biomass indicator. In addition to this, the recent work of Sharma et al. (1977) with fungal cultures has shown that the chitin content of the cell walls varied with age, morphology, nutrient content of the growth medium and dissolved oxygen concentration.

a. Acid and alkaline hydrolysis have been compared by Swift (1973b) for estimates of fungi in plant litter. He concluded that the former, although it was the more involved procedure, gave more consistent yields than the alkaline hydrolysis. Brief details of the acid procedure are included here; those of the alkaline hydrolysis are given by Ride & Drysdale (1972). The alkaline method should be used with caution on freshwater plankton or benthos, because it is likely to extract numerous interfering coloured compounds, including humic materials.

No more than 100 mg of natural material is incubated at room temperature for 3 h in a tube with 5 ml 6 M HCl. The tube is then sealed and hydrolysis completed at 80 °C for 16 h. The hydrolysate is filtered and evaporated at 75 °C and the residue is redissolved in distilled water. The glucosamine concentration at this stage should be in the range 4–14 μg ml^{-1}. The solution is dispersed on to a Dowex 50 (200–400 mesh) cation-exchange column which is washed with water, and the glucosamine is then eluted with 2 M HCl. This solution is neutralized with NaOH and the sugar amine condensed with acetylacetone (freshly prepared 2% in 1 M Na_2CO_3) at 89-92 °C for 45 min. The condensate is mixed with Ehrlich's reagent (2·67% (w/v) *p*-dimethylaminobenzaldehyde in 1+1 ethanol and concentrated HCl). The pink colour which develops is determined spectrophotometrically at 530 nm. The method is relatively insensitive, the detection limit between samples being about 8 μg glucosamine. The procedure as outlined is thought to yield 70–80% of the glucosamine in the original chitin. Swift (1973a, b) emphasizes that the conversion factor from glucosamine to mycelial biomass must be calculated with care, since the chitin content of the mycelium varies with age, condition and the species involved. He comments that the method may overestimate the mycelium biomass, and should only be used on samples where sources of insect chitin are absent.

b. Enzymatic cleavage of chitin yields N-acetylglucosamine, considered to be a more specific indicator of chitin than glucosamine, which is the product of chemical hydrolysis and may also be derived from sources other than the fungal cell-wall polymer (Willoughby 1978). Presumably the presence of a large bacterial biomass (the cell-wall mucopeptides contain N-acetylglucosamine) might interfere with this determination if the cell wall is hydrolysed during the incubation period. The other disadvantage of the method is that the reaction is slow, and a long incubation period may be required to hydrolyse all the substrate. This is usually considered to be the point at which N-acetylglucosamine ceases to accumulate (Willoughby 1978). The method described below is essentially that of Tracey (1955).

Commercial chitinase, or a solution of the enzyme prepared from puffballs (*Lycoperdon* spp.), is mixed with 0·08 M sodium acetate–acetic acid buffer at pH 4·8 (Willoughby 1978). Natural material and test fungal material is autoclaved, before incubation with the enzyme at 33 °C in a stoppered vial. The colour is developed with Ehrlich's reagent (2 g dissolved in 100 ml glacial acetic acid with 5 ml concentrated HCl) as follows:

1 ml of test or standard solution is mixed with 0·3 ml of saturated sodium borate in a tube, which is then capped and placed in a boiling-water bath for 7 min. The tube is cooled, the contents made up to 10 ml with glacial acetic acid and then 1 ml of the Ehrlich reagent is added. The colour is allowed to develop for 45 min after mixing the reagents, before it is read at 540 nm.

Willoughby (1978) observed that chitin hydrolysis could continue for more than 10 days and concluded that the reaction should be allowed to proceed until a constant N-acetylglucosamine figure is reached. The method can provide comparative information and is probably most useful during phases of active growth of the fungi and the environment; i.e. as a biomasss indicator it is probably most useful when the vast bulk of the fungal tissue is living. For comparison of different environmental samples, useful information will only be obtained (as with chemical hydrolysis) if the age, condition and species composition of the fungal population and the extraction efficiency of the method are known for each substratum or environmental sample.

6.4.2.2 *Bacterial biomass*

Almost all prokaryote cell walls contain mucopeptides (mureins, peptidoglycans), the distinguishing features of which include the presence of the amino sugar N-acetylmuramic acid, and often the amino acid diaminopimelic acid. The mucopeptide accounts for *c.* 40–90% of the wall dry weight of Gram-positive bacteria and 5–10% of the wall dry weight of Gram-negative species. Diaminopimelic acids are present in Gram-negative bacteria and actinomycetes but they are replaced by lysine in other forms. There is little information on the proportion of these substances in cyanobacteria. Of the two cell-wall materials, therefore, muramic acid appears to be the more promising, but the method may be used only in samples in which cyanobacterial populations are absent, if bacterial, as opposed to prokaryotic biomass is to be estimated.

a. Enzymatic analysis. An example of the application to sediments of the technique is given by Moriarty (1977), whose method is summarized below.

A sample of sediment or bacterial suspension (200-500 mg or 10-20 mg respectively) is hydrolysed with 3 M HCl (1 ml) in a sealed tube at 100°C for 6 h. Phosphate solution (0.2 ml of 0.5 M Na_2HPO_4) is added, followed by neutralization with 5 M NaOH and clarification by filtration or centrifugation. An aliquot of the sample is stored in this condition for determination of extraneous lactate, because the final determination of the muramic acid is by its conversion to D-lactate. The pH of the sample is raised to 12.5 with NaOH, and it is incubated for 2 h at 35 °C. The pH is then reduced to 8-8.5 with HCl, and the sample is reclarified if necessary. The lactate is converted through pyruvate, using D(-)lactic acid dehydrogenase, E.C. 1.1.1.28 (LDH), to alanine with glutamate pyruvate transaminase E.C. 2.6.1.2. (GPT), and the NADH is assayed with bacterial luciferase (Sigma Chemical Co. Ltd.).

A small aliquot of the hydrolysate is dispensed into a test-tube and the volume is made up to 100 μl with distilled water if necessary. The aliquot should contain 10-200 ng D-lactate. To this is added 0.5 ml of reagent; the contents are mixed rapidly and then incubated at 30 °C for 15 min. The reagent contains the following by proportion – 500 volumes glutamate buffer 0.3 M, pH 9.0 : 10 volumes NAD 33 mg ml^{-1} : 1 volume LDH 5 mg ml^{-1} : 0.5 volumes GPT 10 mg ml^{-1}. After incubation the tubes are transferred to an ice bath. 100 μl of this solution is then added to a phosphate-luciferase preparation and the light emission is measured. This may be done in a scintillation spectrometer for a given period after mixing, or immediately in a suitable photometer (see section 6.3.1 c). A solution of bacterial luciferase (1.5 mg ml^{-1}) is prepared with an equivalent amount of bovine serum albumen in phosphate buffer. This is stored on ice for 15 min and cleared by centrifugation at 3000 g for 5 min at 0 °C. The phosphate preparation contains proportionally 200 volumes phosphate buffer, 0.1 M, pH 7.5 : 14 volumes 2-mercaptoethanol : 10 volumes flavin mononucleotide 0.5 mg ml^{-1} : 10 volumes dodecyl aldehyde (saturated) in ethanol. The dodecyl aldehyde is centrifuged immediately before use. For samples containing 10-50 ng lactate, 100 μl of luciferase preparation is added to 2 ml of the phosphate preparation; for 50-200 ng lactate 50 µl will suffice. It is to this mixture that the 100 µl of lactate–enzyme solution is added. The readings obtained are compared with those of standard solutions of lactate. Other sensitive assays for lactate are available (King & White 1977), and if necessary fluorimetric procedures may be used (Guilbault & Kramer 1964). Under such circumstances, careful checks for interference have to be made, particularly if benthic material is being analysed. It is assumed that all enzyme-catalysed substrate analyses comply with the underlying assumptions of the tests (e.g. linearity of reaction with time, enzyme concentration, etc.).

b. Chemical analysis. King & White (1977) measured muramic acid in estuarine and marine samples and concluded that certain enzymic methods overestimated the concentration of lactate because of non-specificity of the enzymes. They propose the following chemical analysis.

An aliquot of sample (1–5 g) is hydrolysed with 6 M HCl for 4·5 h at 105 °C in a test-tube sealed with a Teflon-lined screwcap. The contents are washed quantitatively through a coarse sintered-glass filter and then dried at 55 °C under reduced pressure. The material is then dissolved in acetone : 0·1 M HCl (9:1 v/v) and separated by cellulose thin-layer chromatography with four cycles of acetone : glacial acetic acid : water (9:1:1 v/v). Material with an R_f range of 0·35–0·7 is eluted with methanol : water (7:3 v/v) concentrated under N_2, 1 ml is dispensed into a test-tube and 0·5 ml of 1 M NaOH added. The mixture is incubated for 30 min at 38 °C, and 10 ml of concentrated H_2SO_4 is added before sealing the test-tube as above and placing it in a boiling water bath for 5 min. After cooling, 0·1 ml of 4% (w/v) $CuSO_4$ and 0·2 ml of 1·5% (w/v) p-hydroxydiphenyl in 95% (v/v) ethanol are added, mixed immediately in the restoppered tube and then incubated at 30 °C for 30 min. The absorbance is read at 560 nm against muramic acid standards. The method is suitable for lactate concentrations of 5 to 20 µg ml^{-1}.

If the proportion of Gram-negative rods and other bacteria is known then the muramic acid value may be converted to a first approximation of bacterial biomass, using the formula (Moriarty 1977)

$$C = \frac{1000M}{12n+40p}$$

where C is carbon (mg), M is muramic acid (µg) and $n+p = 1$ where n is the proportion of Gram-negative and weakly Gram-positive bacteria and p is the proportion of strongly Gram-positive bacteria. If the Gram reaction of the population is to be determined, this is best done directly on the natural material. Concentrates of bacterioplankton and samples of benthos should be stained directly rather than relying on estimates from the Gram reaction of isolates obtained by enrichment or viable count techniques.

Alternatively, an average value for the muramic acid content of the bacteria may be assumed. Moriarty (1977) concluded that if most of the benthic bacteria were Gram-negative or weakly Gram-positive, an average value of 12 µg muramic acid per mg of bacterial C was applicable. King & White (1977) obtained a value of 6·4 µg muramic acid mg^{-1} of bacteria, but also noted that blue-green algae could contain 500 times more than this. The assay is therefore likely to provide information on prokaryotic rather than bacterial biomass.

Because they are absent entirely from some groups of bacteria, diaminopimelic acids are less satisfactory as biomass indicators. If, however, they are to be assayed, the extremely sensitive method developed by Rogers, Chambers & Clarke (1967) should be considered. The diaminopimelic acid is mixed with phthalaldehyde to yield a highly fluorescent complex ($\lambda_{ex} = 365$ nm, $\lambda_{em} = 425$ nm). The method is 10^4 times more sensitive than the best colorimetric procedure with a detection limit of 9 ng.

6.4.3 *Other methods for bacteria*

The two methods described in this section are relatively new and their application to natural materials requires further investigation. Reports on their use are however appearing with increasing frequency (particularly 6.4.3.1) and therefore brief summaries are provided here.

6.4.3.1 *Lipopolysaccharide assay*

The aqueous extract of amoebocytes from *Limulus polyphemus* (the horseshoe crab) will form a gel or turbid solution in the presence of minute quantities of the protein–lipopolysaccharide complex endotoxin from Gram-negative bacteria. The optical density of the mixture is related to the density of the bacterial populations. The endotoxin occurs only in the cell walls of Gram-negative bacteria which are thought to predominate in the aquatic environment, and appears to be related to biomass rather than cell numbers with a reported C : endotoxin ratio of 6.35 (although a good correlation with bacterial epifluorescence counts (see section 3.4.3.) was also obtained from a large number of marine samples (Watson et al. 1977)).

Considerable care is necessary to reduce background levels during this assay. Ignited (550 °C overnight) or dry-heat-sterilized (170 °C for 1 h) glassware or disposable tubes should be used, and all solutions should be made up in sterile, pyrogen-free distilled water. The endotoxin preparation is supplied in sealed vials and is reconstituted in pyrogen-free water immediately before use. The sample and endotoxin are then mixed and incubated for 1 h at 37 °C. After incubation the contents are mixed gently and the absorbance is read at 360 nm. Suitable controls are also tested and the corrected absorbance is compared with those obtained with standard endotoxin solution. Dilution of the sample in a solution of NaCl is known to enhance the sensitivity of the assay but 10^{-5} M EDTA in 0·9% NaCl is even more effective (Coates 1977) and also reduces the variation between bacterial strains. It is thus possible to detect down to 500 bacteria per sample.

6.4.3.2 *Luminol chemiluminescence*

The method is based on that of Oleniacz et al. (1968). The luminescence is caused by the stimulation of H_2O_2 by heme proteins in bacteria. The peroxide reacts with the alkaline luminol (5-amino-2, 3-dihydro-1, 4-phthalazinedione), causing light emission.

The reagents are prepared as follows: 0·01 g luminol is dissolved in 2 ml 0·2 M NaOH, 0·36 g glucose is added and the volume made up to 100 ml with distilled water. Two volumes of this mixture are added to 1 volume of 1·5% (w/v) $NaBO_3.4H_2O$ (sodium perborate) in distilled water and 2 volumes of 0·2 M NaOH, just before use. To 9 ml of water sample (or a 0·2 μm polycarbonate membrane-filtered control) is added 1 ml of 2 M NaOH. A known volume of the alkaline sample is placed in a tube in a suitable photometer (see section 6.3.1.c) and an equal volume of the reagent is injected with a syringe. The light emitted is integrated over a period of 60 s. It is necessary to add alkali to the samples to reduce the high background values of the controls, particularly those of acidic material.

A number of solutes (particularly iron) also induce chemiluminescence, and although membrane-filtered controls may correct for this to some degree, the method may be unsatisfactory for anoxic samples. Although it has been claimed that a detection limit of 10^3 bacteria ml^{-1} can be achieved, a detailed analysis by Miller & Vogelhut (1978) demonstrated that this was likely to be closer to 10^5–10^6 bacteria ml^{-1}. This, combined with potential interferences, limits the usefulness of the technique in fresh water.

6.5 ANALYSIS OF RESULTS

If internal standards are used whenever possible, particularly when interfering substances are present, then it is more likely that quantitative estimates will be obtained with the chemical/biochemical techniques for biomass than with counting methods for population estimates.

It is also generally accepted that population estimates contain error terms and that problems sometimes exist in the calculation of reliable confidence limits. Biomass estimates, usually obtained by biochemical or chemical analysis are, unfortunately, rarely subjected to the same rigorous statistical treatment. Yet these results may fail to conform to the assumptions underlying parametric statistical tests (see Appendix A), particularly with regard to the relationships between data variance and the mean concentration of the biomass indicator.

It is not uncommon for papers on biomass analyses to contain a statement of the coefficient of variation (occasionally without reference to sample size and mean value) and the lower limit of detection (often without a definition of the latter). The reader is in a far better position to judge the performance of a particular method if provided with the following information:

(a) the standard deviation or coefficient of variation for a given mean value and sample size;

(b) an assessment of the variability of the data over the range of levels recorded;

(c) the detection limit, which is defined as the value by which the sample must exceed the blank to be statistically valid. This value is given by

$$L_d = 2\sqrt{2} \cdot t \cdot S_b \quad \text{(Wilson 1973)}$$

where L_d = the detection limit,

t = the tabulated value of Student's t (one-tailed) for $f = n - 1$ degrees of freedom and $P = 0.05$

S_b = the estimated within-batch standard deviation (at f degrees of freedom) of the blank (in concentration units)

This is approximated by $L_d = 4.65\ S_b$ for $f > 50$

The term 'detection limit' is preferred to 'sensitivity', although they are often used interchangeably. The latter should be used to define machine response per unit of substance being analysed.

7. ACKNOWLEDGMENTS

I am extremely grateful to the following who have contributed to this volume in a very positive way: Dr J. H. Baker, Mr E. D. Le Cren, Dr S. R. Elsden, Mirna J. L. G. Orlandi, Dr S. I. Heaney, Mr J. E. M. Horne, Drs J. M. Shewan, D. W. Sutcliffe, J. F. Talling, B. A. Whitton, P. J. Le B. Williams and L. G. Willoughby, who have read and commented on all or part of this manuscript; colleagues in the Freshwater Biological Association and the Aquatic Microbiology Discussion Group, who have provided the stimulus of critical discussion and made much of this volume possible; Mr B. M. Simon who, over eight years, has contributed, with characteristic care and attention to detail, to the testing and re-testing of methods discussed in this book; Mr T. I. Furnass for preparing the figures; my wife, Helen, for her constant help and encouragement throughout the many stages and versions of this publication, and for all the effort which went into the preparation of the index.

8. APPENDICES

Appendix A: A summary guide to statistical methodology

This is not a statistical text but merely a suggested approach to the analysis of data. Details of the methods and tests included may be found in the references provided. If no reference is given, details, usually with a worked example, are to be found in Elliott (1977).

The notation used follows the usual statistical conventions.

Statistic	Notation
a variate	Y_i
sample size	n
mean	$\bar{Y} = \Sigma Y_i / n$
deviation (from the mean)	$y = Y_i - \bar{Y}$
sums of squares	$\Sigma y^2 = \Sigma (Y_i - \bar{Y})^2$
variance	$S^2 = \Sigma y^2 / (n-1)$
standard deviation	$S = \sqrt{S^2}$
standard error (of the mean)	$SE = \sqrt{S^2/n}$
ceofficient of variation (%)	$CV = (S/\bar{Y}) \times 100$

Experimental design and the statistical analysis of data may be divided into six stages.

1. Statement of null hypothesis

This is a specific hypothesis about the samples or populations. It is usually a statement of no difference (hence *null* hypothesis), that the samples are from the same population and are within the accepted error of the population mean.

It is essential to state clearly the objectives of the experiment or sampling programme, and to translate these objectives into precise questions which may be tested as a null hypothesis.

2. Specification of the significance level and the number of samples required

The significance level to which one wishes to work may require a degree of replication which is impractical. For most biological data the 95% level is acceptable, i.e. a probability of 1 in 20 that the results are due to chance. The number of samples required is inextricably bound to the significance level chosen. A simple method for estimating the number of samples or replicates (n) is given by the formula

$$n = \frac{S^2}{D^2 \bar{Y}^2}$$

where $\bar{Y}$ and S^2 are the mean and variance of the variable (usually derived from earlier data) and D is an expression of the standard error as a proportion of the mean at the level of significance desired; e.g. if a standard error of 20% of the mean is required then $D = 20/100 = 0 \cdot 2$.

In the absence of information on the variance of the sample, a simple rule of thumb may be adopted. The sample size is increased in units of 5 (i.e. $n = 5, 10, 15$ etc) and at each stage $\bar{Y}$ is calculated. When $\bar{Y}$ no longer fluctuates then a satisfactory number of sampling units has been achieved.

These tests are worth performing, if only to illustrate how far one's own estimate of the required degree of replication falls short of reality. It is worth emphasizing at this stage that many published statistical tests are for large samples (i.e. $n > 50$), a criterion which many ecological data sets do not satisfy.

3. Specification of the dimensions and location of the sampling unit

The dimensions of the sampling unit may be controlled to some degree by the number of samples required, but the major constraint is usually the nature of the problem posed. For example the measurement of sediment respiration might require a sample area of no less than, say, 25 cm². A detailed study of the micro-distribution of ciliate/bacterial relationships, on the other hand, might be conducted on areas of less than 1 cm². It is usually more efficient to have more small sampling units than a few large ones, the statistical errors being thus reduced. This may create sample handling problems. Many microbiologists take many small samples at the site and then combine these, rather than taking one large sample. This, of course, leads to a loss of information, but such compromises may be necessary. The researcher should be aware of the problems of sampling, pooling, sub-sampling, and the magnitude of the errors introduced at each stage. A nested analysis of variance will provide information on the size of these errors and help to optimize sampling procedure.

The location of the sampling unit will depend on the nature of the problem. Studies of population distribution theoretically require that samples should be taken in a random manner. Three-dimensional sampling in a lake or river in such a way, however, may be very difficult to achieve. Where information relating the population to time or positional changes is required, random sampling may not be suitable, and stratified random or systematic sampling is employed. Although no worker would ever assume vertical homogeneity in a stratified lake, the assumption of horizontal homogeneity is often made, in spite of the existence of a large and growing body of evidence to the contrary.

4. *Determination of the sample distribution, the data transformation required, and calculation of confidence limits*

Conventional parametric statistical tests assume that data conform to a normal distribution. If this condition is met, it usually follows that the variance is independent of the mean, that the variances are homogeneous and that the components of variance are additive. This section outlines some tests for checking these assumptions.

4.1 *Determination of the distribution of the data*

Data may be tested against known distributions as follows:

Distribution	*Test*
(a) Random	χ^2, S^2 to $\bar{Y}$ ratio ($n<31$)
	χ^2 test for goodness of fit ($n>31$)
	G test, Sokal & Rohlf (1969)
(b) Contagious	U and T statistics ($n<50$)
	χ^2 test for goodness of fit ($n>50$)
	G test, Sokal & Rohlf (1969)
(c) Regular	χ^2, S^2 to $\bar{Y}$ ratio
(d) Normal	W test (n 20-50), Hahn & Shapiro (1967)
	Lilliefors test ($n>4$), Conover (1971)
	Kolmogorov-Smirnov test, Conover (1971)
	Graphical method using arithmetic probability paper ($n>50$) or rankits ($n<50$), Sokal & Rohlf (1969).
(e) Unknown distribution	Taylor's Power Law
(f) Continuous sample distribution	Kolmogorov-Smirnov test, Sokal & Rohlf (1969)

4.2 *The choice of transformation* will depend on the parent distribution. The following transformations are commonly used:

Distribution	Characteristic	Transformation
Random	$S^2 = \bar{Y}$	$\sqrt{Y}$ or $\sqrt{Y+0.5}$
Contagious	$S^2 > \bar{Y}$	$\log Y$ or $\log(Y+1)$
Regular	$S^2 < \bar{Y}$	Y^2
Unknown	$S^2 \propto \bar{Y}$	Y^x (Taylor's Power Law)
Data expressed as a percentage or proportion, p		$\arcsin\sqrt{p}$

4.3 *The adequacy of the transformation* must then be checked either by
(a) Graphical test, plotting S^2 on $\bar{Y}$ (double log plot) or
(b) Checking the homogeneity of the variances by calculating S^2_{max}/S^2_{min} or by Bartlett's test.

4.4 *Calculation of confidence limits*
It is always desirable to give an estimate of the precision of a mean value, and 95% confidence limits may be calculated from the following formulae:

Condition	Formula
Large samples ($n > 30$)	$\bar{Y} \pm t\sqrt{\dfrac{S^2}{n}}$
Small samples ($n < 30$) (a) Random distribution $n\bar{Y} > 30$	$\bar{Y} \pm t\sqrt{\dfrac{\bar{Y}}{n}}$
$n\bar{Y} < 30$ or single counts	from tables or $Y_i \pm 2\sqrt{Y_i}$
(b) Transformed data	Formula as for large samples but transformed data used.
(c) MPN counts	See section 5.3.3.

5. *Design of the experiment (or sampling programme) and choice of the statistical test*

When designing an experiment, it is often useful to put on paper the number of samples, treatments, replicates, etc intended. The total number of experimental tubes, flasks or samples required will then be the product of all these factors. Wherever necessary include control treatments in the experimental design. The statistical test chosen will depend on a number of factors, including the number of variables, and whether a comparison between values is being made, or the mathematical relationship between the variables is being determined. If the data conform to a normal distribution, or can be transformed to satisfy the conditions outlined above, then parametric statistical tests may be used (e.g. t-test, analysis of variance). Details of these tests may be obtained from standard works: Elliott (1977), Snedecor & Cochran (1967), Sokal & Rohlf (1969). If attempts to analyse and transform the data fail, or if insufficient data are available, then non-parametric tests may be used. These tests do not assume a particular underlying distribution but usually require that the data sets have the same (or similar) distributions. Details of these tests may be found in Siegel (1956), Campbell (1974) and Conover (1971). Further discussion on the application of non-parametric tests to microbiological data, with worked examples, is provided by Jones (1973b), and tests for MPN results are given in section 5.3.3.

6. *Compute the test and draw conclusions*

And so on to the next Null Hypothesis.

Appendix B: List of Manufacturers and Suppliers.

Equipment and materials not listed here may be obtained from the major laboratory suppliers.

Chapter 2

Ultra-Turrax homogeniser	Sartorius Instruments Ltd, 18 Avenue Rd, Belmont, Surrey.
Ultrasonic disintegrator	MSE Scientific Instruments, Manor Royal, Crawley West Sussex.

Chapter 3

Membrane filters: cellulosic and apparatus	Millipore U.K. Ltd, "Millipore House", Abbey Rd, London NW10 7SP.
	Sartorius, V. A. Howe & Co. Ltd, 88 Peterborough Rd, London SW6 3EP.
	Gelman Hawksley Ltd, 10 Harrowden Rd, Brackmills, Northampton NN4 0EB.
	Schleicher & Schuell Selectron Filters, Anderman & Co. Ltd, Central Avenue, East Molesey, Surrey KT8 0QT.
Membrane filters: polycarbonate	Nuclepore, Sterilin Ltd, 43-45 Broad St, Teddington, Middlesex TW11 8QT.
	Uni-Pore, Bio-Rad Laboratories Ltd, Caxton Way, Holywell Industrial Estate, Watford, Hertfordshire WD1 8RP.
Glass-fibre filters	Whatman Labsales Ltd, Springfield Mill, Maidstone, Kent ME14 2LE.
	Gelman Hawksley Ltd. Address above.
Nylon, metal and other filters	John Stanier & Co. Manchester Wire Works, Sherbourne St, Manchester M3 1FD.
Micromesh	EMI, Electron Tube Division, 243 Blyth Rd, Hayes, Middlesex UB3 1HJ.
Swin-Lok® filter holders	Sterilin Ltd. Address above.
Dylon	Hardware stores or Dylon International Ltd, Dylon Works, Sydenham, London SE26.
Piston pipette	Finnpipette, Jencons Ltd, Mark Rd, Hemel Hempstead, Hertfordshire HP2 7DE.

APPENDIX B

Chapter 4

Glass solder (X76)	Glass Tubes Components Ltd, Glass Works, Sheffield Rd, Chesterfield.
Lund Chamber	Water Research Centre, P.O. Box 16, Ferry Lane, Medmenham, Marlow, Buckinghamshire SL7 2HD.
Perfilev & Gabe Capillaries	Camlab Ltd, Nuffield Road, Cambridge CB4 1TH.

Chapter 5

Plastic micro-titre dishes	Sterilin Ltd. Address above.

Chapter 6

Luciferin – luciferase preparations	Sigma London Chemical Co. Ltd, Fancy Rd, Poole, Dorset BH17 7NH.
Luciferin	Sigma London Chemical Co. Ltd. Address above.
	Calbiochem Ltd, 78/81 South St, Bishop's Stortford, Hertfordshire CM23 3AL.
Aminco Chem-Glow Photometer and Hamilton constant-rate adjustable syringe	V. A. Howe Ltd, 88 Peterborough Rd, London SW6 3EP.
Field Fluorimeter	Turner III V. A. Howe Ltd. Address above.
Limulus extract and endotoxin standards	Worthington Biochemical Corporation, (Pyrostat™) Cambrian Chemicals Ltd, George St, Croydon.

9. REFERENCES

References marked * are general works which may not necessarily be cited in the text.

*Aaronson, S. (1970). *Experimental microbial ecology.* New York. Academic Press.
*American Public Health Association (1976). *Standard methods for the examination of water and waste water.* 14th edn, Washington, American Public Health Association.
Atkinson, D. E. & Walton, G. M. (1967). Adenosine triphosphate conservation in metabolic regulation. *J. biol. Chem.* **193**, 265-275.
Babuik, L. A. & Paul, E. A. (1970). The use of fluorescein isothiocyanate in the determination of the bacterial biomass of grassland soil. *Can. J. Microbiol.* **16**, 57-62.
Badger, E. H. M. & Pankhurst, E. S. (1960). Experiments on the accuracy of surface drop bacterial counts. *J. appl. Bact.* **23**, 28-36.
Baker, J. H. & Farr, I. S. (1977). Origins, characterization and dynamics of suspended bacteria in two chalk streams. *Arch. Hydrobiol.* **80**, 308-326.
Banse, K. (1977). Determining the carbon to chlorophyll ratio of natural phytoplankton. *Mar. Biol.* **41**, 199-212.
Belly, R. T., Bohlool, B. B. & Brock, T. D. (1973). The genus *Thermoplasma*. *Ann. N.Y. Acad. Sci.* **225**, 94-107.
Bochner, B. R. & Savageau, M. A. (1977). Generalized indicator plate for genetic, metabolic, and taxonomic studies with microorganisms. *Appl. environ. Microbiol.* **33**, 434-444.
Bohlool, B. B. & Brock, T. D. (1974). Population ecology of *Sulfolobus acidocaldarius*. II. Immunological studies. *Arch. Mikrobiol.* **97**, 181-194.
Bohlool, B. B. & Schmidt, E. L. (1968). Non-specific staining: its control in immunofluorescence examination of soil. *Science, N.Y.* **162**, 1012-1014.
Bonde, G. J. (1977). Bacterial indication of water pollution. In *Advances in aquatic microbiology* (ed. M. R. Droop and H. W. Jannasch) **1**, 273-364. London. Academic Press.
Bousefield, I. J., Smith, G. L. & Trueman, R. W. (1973). The use of semi-automatic pipettes in the viable counting of bacteria. *J. appl. Bact.* **36**, 297-299.
Bowden, W. B. (1977). Comparison of two direct-count techniques for enumerating aquatic bacteria. *Appl. environ. Microbiol.* **33**, 1229-1232.

Bowie, I. S. & Gillespie, P. A. (1976). Microbial parameters and trophic status of ten New Zealand lakes. *N.Z.Jl mar. Freshwat. Res.* **10**, 343-354.

Brock, M. L. & Brock, T. D. (1968). The application of micro-autoradiographic techniques to ecological studies. *Mitt. int. Verein. theor. angew. Limnol.* **15**, 1-29.

Caldwell, D. E. (1977). Accessory pigment fluorescence for quantitation of photosynthetic microbial populations. *Can. J. Microbiol.* **23**, 1594-1597.

Caldwell, D. E. & Tiedje, J. M. (1975a). A morphological study of anaerobic bacteria from the hypolimnia of two Michigan lakes. *Can. J. Microbiol.* **21**, 362-376.

Caldwell, D. E. & Tiedje, J. M. (1975b). The structure of anaerobic bacterial communities in the hypolimnia of several Michigan lakes. *Can. J. Microbiol.* **21**, 377-385.

*Campbell, R. C. (1974). *Statistics for biologists*. Cambridge. University Press.

Cassell, E. A. (1965). Rapid graphical method for estimating the precision of direct microscopic counting data. *Appl. Microbiol.* **13**, 293-296.

Cattolico, R. A. & Gibbs, S. P. (1975). Rapid filter method for the microfluorimetric analysis of DNA. *Analyt. Biochem.* **69**, 572-582.

Cavari, B. (1976). ATP in Lake Kinneret: indicator of microbial biomass or of phosphorus deficiency. *Limnol. Oceanogr.* **21**, 231-236.

Chapman, A. G., Fall, L. & Atkinson, D. E. (1971). Adenylate energy charge in *Escherichia coli* during growth and starvation. *J. Bact.* **108**, 1072-1086.

Chu, S. P. (1942). The influence of the mineral composition of the medium on the growth of planktonic algae. 1. Methods and culture media. *J. Ecol.* **30**, 284-325.

Coates, D. A. (1977). Enhancement of the sensitivity of the *Limulus* assay for the detection of Gram-negative bacteria. *J. appl. Bact.* **42**, 445-449.

Cochran, W. G. (1950). Estimation of bacterial densities by means of the "most probable number". *Biometrics,* **2**, 105-116.

*Collins, V. G., Jones, J. G., Hendrie, M. S., Shewan, J. M., Wynn-Williams, D. D. & Rhodes, M. E. (1973). Sampling and estimation of bacterial populations in the aquatic environment. In *Sampling – microbiological monitoring of environments* (ed. R. G. Board and D. W. Lovelock). *Soc. appl. Bact. tech. Ser.* No. 7, 77-110. London. Academic Press.

Collins, V. G. & Kipling, C. (1957). The enumeration of waterborne bacteria by a new direct count method. *J. appl. Bact.* **20**, 257-264.

Collins, V. G. & Willoughby, L. G. (1962). The distribution of bacteria and fungal spores in Blelham Tarn with particular reference to an experimental overturn. *Arch. Mikrobiol.* **43**, 294-307.

*Conover, W. J. (1971). *Practical nonparametric statistics*. New York. John Wiley.

Cruickshank, C. N. D., Cooper, J. R. & Conran, M. B. (1959). A new tissue culture chamber. *Exp. Cell Res.* **16**, 695-698.

*Curds, C. R. (1969). An illustrated key to the British freshwater ciliated Protozoa commonly found in activated sludge. *Tech. Pap. Wat. Pollut. Res.* No. 12. London. HMSO.

Curds, C. R. & Bazin, M. J. (1977). Protozoan predation in batch and continuous culture. In *Advances in aquatic microbiology* (ed. M. R. Droop and H. W. Jannasch) **1**, 115-176. London. Academic Press.

Curds, C. R., Roberts, D. M. & Wu Chih-Hua (1978). The use of continuous cultures and electronic sizing devices to study the growth of two species of ciliated Protozoa. In *Techniques for the study of mixed populations* (ed. D. W. Lovelock & R. Davies). *Soc. appl. Bact. tech. Ser.* No. 11, 165-177. London. Academic Press.

Dodson, A. N. & Thomas, W. H. (1964). Concentrating plankton in a gentle fashion. *Limnol. Oceanogr.* **9**, 455-456.

Dozier, B. J. & Richerson, P. J. (1975). An improved membrane filter method for the enumeration of phytoplankton. *Verh. int. Verein. theor. angew. Limnol.* **19**, 1524-1529.

Drake, J. F. & Tsuchiya, H. M. (1973). Differential counting in mixed cultures with Coulter counters. *Appl. Microbiol.* **26**, 9-13.

Dring, D. M. (1971). Techniques for microscopic preparation. In *Methods in microbiology*, **4** (ed. C. Booth) 95-111. London. Academic Press.

Eaton, J. W. & Moss, B. (1966). The estimation of numbers and pigment content in epipelic algal populations. *Limnol. Oceanogr.* **11**, 584-595.

Edmondson, W. T. (1974). A simplified method for counting phytoplankton. In *A manual on methods for measuring primary production in aquatic environments* (ed. R. A. Vollenweider). IBP Handbook No. 12, 2nd edn, 14-16. Oxford. Blackwell Scientific Publications.

Edwards, S. W. & Lloyd, D. (1977). Changes in oxygen uptake rates, enzyme activities, cytochrome amounts and adenine nucleotide pool levels during growth of *Acanthamoeba castellani* in batch culture. *J. gen. Microbiol.* **102**, 135-144.

*Elliott, J. M. (1977). Some methods for the statistical analysis of samples of benthic invertebrates. *Scient. Publs Freshwat. biol. Ass.* No. 25, 2nd edn.

Evans, J. H. (1972). A modified sedimentation system for counting algae with an inverted microscope. *Hydrobiologia*, **40**, 247-250.

Fáust, M. A. & Correll, D. L. (1977). Autoradiographic study to detect metabolically active phytoplankton and bacteria in Rhode River Estuary. *Mar. Biol.* **41**, 293-305.

Finlay, B. J., Laybourn, J. & Strachan, I. (1979). A technique for the enumeration of benthic ciliated protozoa. *Oecologia*, **39**, 375-377.

Fisher, R. A., Thornton, H. G. & Mackenzie, W. A. (1922). The accuracy of the plating method of estimating the density of bacterial populations. *Ann. appl. Biol.* **9**, 325-359.

*Fisher, R. A. & Yates, F. (1948). *Statistical tables for biological, agricultural and medical research.* 3rd edn. Table VIII2. Edinburgh. Oliver and Boyd.

Fliermans, C. B. & Schmidt, E. L. (1975). Autoradiography and immunofluorescence combined for autecological study of single cell activity with *Nitrobacter* as a model system. *Appl. Microbiol.* 30, 674-677.

Fry, J. C. & Humphrey, N. C. B. (1978). Techniques for the study of bacteria epiphytic on aquatic macrophytes. In *Techniques for the study of mixed populations* (ed. D. W. Lovelock & R. Davies). *Soc. appl. Bact. tech. Ser.* No. 11, 1-29. London. Academic Press.

Gaertner, A. (1968). Eine Methode des quantitativen Nachweises niederer, mit Pollen köderbarer Pilze im Meerwasser und im Sediment. *Veröff. Inst. Meeresforsch. Bremerh.*, Sonderbd. 3, 75-91.

*Golterman, H. L., Clymo, R. S. & Ohnstad, M. A. M. (1978). *Methods for physical and chemical analysis of fresh waters.* IBP Handbook No. 8 (2nd edn). Oxford. Blackwell Scientific Publications.

Gorham, P. R., McLachlan, J., Hammer, U. T. & Kim, W. K. (1964). Isolation and culture of toxic strains of *Anabaena flos-aquae* (Lyngb.) de Bréb. *Verh. int. Verein. theor. angew. Limnol.* 15, 796-804.

Goulder, R. (1971). Vertical distribution of some ciliated Protozoa in two freshwater sediments. *Oikos,* 22, 199-203.

Goulder, R. (1974). The seasonal and spatial distribution of some benthic ciliated Protozoa in Esthwaite Water. *Freshwat. Biol.* 4, 127-147.

Goulder, R. (1975). The effects of photosynthetically raised pH and light on some ciliated Protozoa in a eutrophic pond. *Freshwat. Biol.* 5, 313-322.

Green, B. L., Clausen, E. & Litsky, W. (1975). Comparison of the new Millipore HC with conventional membrane filters for the enumeration of faecal coliform bacteria. *Appl. Microbiol.* 30, 697-699.

Guilbault, G. G. & Kramer, D. N. (1964). New fluorometric method for measuring dehydrogenase activity. *Analyt. Chem.* 36, 2497-98.

Hahn, G. J. & Shapiro, S. S. (1967). Tests for distribution assumptions. In *Statistical models in engineering,* 294-308. New York. John Wiley.

Halvorson, H. O. & Ziegler, N. R. (1933). Application of statistics to problems in bacteriology. 1. A means of determining bacterial population by the dilution method. *J. Bact.* 25, 101-121.

Harris, J. E., McKee, T. R., Wilson, R. C. & Whitehouse, U. G. (1972). Preparation of membrane filter samples for direct examination with an electron microscope. *Limnol. Oceanogr.* 17, 784-787.

Harris, R. F. & Sommers, L. E. (1968). Plate-dilution frequency technique for assay of microbial ecology. *Appl. Microbiol.* 16, 330-334.

*Hartman, P. A. (1968). *Miniaturized microbiological methods.* New York. Academic Press.

Heaney, S. I. (1978). Some observations on the use of *in vivo* fluorescence technique to determine chlorophyll *a* in natural populations and cultures of freshwater phytoplankton. *Freshwat. Biol.* 8, 115-126.

Herbert, D., Phipps, P. J. & Strange, R. E. (1971). Chemical analysis of microbial cells. In *Methods in microbiology,* 5B, (ed. J. R. Norris and D. W. Ribbons), 209-344. London. Academic Press.

Hobbie, J. E., Daley, R. J. & Jasper, S. (1977). Use of Nuclepore filters for counting bacteria by fluorescence microscopy. *Appl. environ. Microbiol.* 33, 1225-1228.

Hobbie, J. E., Holm-Hansen, O., Packard, T. T., Pomeroy, L. R., Sheldon, R. W., Thomas, J. P. & Wiebe, W. J. (1972). A study of the distribution and activity of micro-organisms in ocean water. *Limnol. Oceanogr.* 17, 544-555.

Holm-Hansen, O. (1973). The use of ATP determinations in ecological studies. *Bull. Ecol. Res. Comm. (Stockholm)*, 17, 215-222.

Holm-Hansen, O. & Booth, C. R. (1966). The measurement of adenosine triphosphate in the ocean and its ecological significance. *Limnol. Oceanogr.* 11, 510-519.

Holm-Hansen, O. & Riemann, B. (1978). Chlorophyll *a* determination: improvements in methodology. *Oikos*, 30, 438-447.

Holm-Hansen, O., Sutcliffe, W. H. & Sharpe, J. (1968). Measurement of deoxyribonucleic acid in the ocean and its ecological significance. *Limnol. Oceanogr.* 13, 507-514.

Hoppe, H.-G. (1976). Determination and properties of actively metabolising heterotrophic bacteria in the sea, investigated by means of microautoradiography. *Mar. Biol.* 36, 291-302.

Hussainy, S. U. (1972). Bacterial and algal chlorophyll in two salt lakes in Victoria, Australia. *Wat. Res.* 6, 1361-1367.

Hutchinson, G. E. (1975). *A treatise on limnology*, Vol. 3; *limnological botany*. New York. Wiley.

Iturriaga, R. & Hoppe, H.-G. (1977). Observations of heterotrophic activity on photoassimilated organic matter. *Mar. Biol.* 40, 101-108.

Iturriaga, R. & Rheinheimer, G. (1975). A simple method for counting bacteria with active electron transport systems in water and sediment samples. *Kieler Meeresforsch.* 31, 83-86.

Jannasch, H. W. (1958). Studies on planktonic bacteria by means of a direct membrane filter method. *J. gen. Microbiol.* 18, 609-620.

Johnston, D. W. & Cross, T. (1976). Actinomycetes in lake muds: dormant spores or metabolically active mycelium? *Freshwat. Biol.* 6, 465-470.

Jones, E. B. G. (1971). Aquatic Fungi. In *Methods in microbiology*, (ed. C. Booth), 4, 335-365. London. Academic Press.

Jones, J. G. (1970). Studies on freshwater bacteria: effect of medium composition and method on estimates of bacterial population. *J. appl. Bact.* 33, 679-687.

Jones, J. G. (1973a). Studies on freshwater bacteria: the effect of enclosure in large experimental tubes. *J. appl. Bact.* 36, 445-456.

Jones, J. G. (1973b). Use of nonparametric tests for the analysis of data obtained from preliminary surveys: a review. *J. appl. Bact.* 36, 197-210.

Jones, J. G. (1974). A method for observation and enumeration of epilithic algae directly on the surface of stones. *Oecologia*, 16, 1-8.

Jones, J. G. (1977). The distribution of some freshwater planktonic bacteria in two stratified eutrophic lakes. *Freshwat. Biol.* 8, 127-140.

Jones, J. G. & Simon, B. M. (1975a). An investigation of errors in direct counts of aquatic bacteria by epifluorescence microscopy, with reference to a new method for dyeing membrane filters. *J. appl. Bact.* **39**, 317-329.

Jones, J. G. & Simon, B. M. (1975b). Some observations on the fluorimetric determination of glucose in fresh water. *Limnol. Oceanogr.* **20**, 882-887.

Jones, J. G. & Simon, B. M. (1977). Increased sensitivity in the measurement of ATP in freshwater samples with a comment on the adverse effect of membrane filtration. *Freshwat. Biol.* **7**, 253-260.

Jones, P. C. T., Mollison, J. E. & Quenouille, M. H. (1948). A technique for the quantitative estimation of soil micro-organisms. *J. gen. Microbiol.* **2**, 54-69.

Jonge, V. N. de & Bouwman, L. A. (1977). A simple density separation technique for quantitative isolation of meiobenthos using colloidal silica Ludox-TM. *Mar. Biol.* **42**, 143-148.

Kapuscinski, J. & Skoczylas, B. (1977). Simple and rapid fluorimetric method for DNA microassay. *Analyt. Biochem.* **83**, 252-257.

Karl, D. M. (1978). Occurrence and ecological significance of GTP in the ocean and in microbial cells. *Appl. environ. Microbiol.* **36**, 349-355.

Karl, D. M. & Holm-Hansen, O. (1976). Effects of luciferin concentration on the quantitative assay of ATP using crude luciferase preparations. *Analyt. Biochem.* **75**, 100-112.

Karl, D. M. & Holm-Hansen, O. (1978). Methodology and measurement of adenylate energy charge ratios in environmental samples. *Mar. Biol.* **48**, 185-197.

Kieffer, D. A. (1973a). Fluorescence properties of natural phytoplankton populations. *Mar. Biol.* **22**, 263-269.

Kieffer, D. A. (1973b). Chlorophyll *a* fluorescence in marine centric diatoms: responses of chloroplasts to light and nutrient stress. *Mar. Biol.* **23**, 39-46.

King, J. D. & White, D. C. (1977). Muramic acid as a measure of microbial biomass in estuarine and marine samples. *Appl. environ. Microbiol.* **33**, 777-783.

Kissane, J. M. & Robins, E. (1958). The fluorometric measurement of deoxyribonucleic acid in animal tissues with special reference to the central nervous system. *J. biol. Chem.* **233**, 184-186.

Knoechel, R. & Kalff, J. (1976). Track autoradiography: a method for the determination of phytoplankton species productivity. *Limnol. Oceanogr.* **21**, 590-596.

Knowles, C. J. (1977). Microbial metabolic regulation by adenine nucleotide pools. In *Microbial energetics*. Symp Soc. gen. Microbiol. No. 27, 241-283. Cambridge. University Press.

Kogure, K., Simidu, U. & Taga, N. (1979). A tentative direct microscopic method for counting living marine bacteria. *Can. J. Microbiol.* **25**, 415-420.

Kubitschek, H. E. (1969). Counting and sizing micro-organisms with the Coulter counter. In *Methods in microbiology,* **1** (ed. J. R. Norris & D. W. Ribbons) 593-610. London. Academic Press.

Lien, T. & Knutsen, G. (1976). Fluorometric determination of DNA in *Chlamydomonas*. *Analyt. Biochem.* **74**, 560-566.

Lin, S. D. (1976). Evaluation of Millipore HA and HC membrane filters for the enumeration of indicator bacteria. *Appl. environ. Microbiol.* 32, 300-302.

Loftus, M. E. & Seliger, H. H. (1975). Some limitations of *in vivo* fluorescence technique. *Chesapeake Sci.* 16, 79-92.

Loftus, M. E. & Carpenter, J. H. (1971). A fluorometric method for determining chlorophylls *a, b* and *c. J. mar. Res.* 29, 319-338.

Lorenzen, C. J. (1966). A method for the continuous measurement of *in vivo* chlorophyll concentration. *Deep-Sea Res.* 13, 223-227.

Lund, J. W. G. (1959). A simple counting chamber for nannoplankton. *Limnol. Oceanogr.* 4, 57-65.

Lund, J. W. G., Kipling, C. & Le Cren, E. D. (1958). The inverted microscope method of estimating algal numbers and the statistical basis of estimations by counting. *Hydrobiologia,* 11, 143-170.

Lund, J. W. G. & Talling, J. F. (1957). Botanical limnological methods with special reference to algae. *Bot. Rev.* 23, 489-583.

*Mackereth, F. J. H., Heron, J. & Talling, J. F. (1978). Water analysis: some revised methods for limnologists. *Scient. Publs Freshwat. biol. Ass.* No. 36, 120 pp.

Madigan, M. T. & Brock, T. D. (1976). Quantitative estimation of bacteriochlorophyll *c* in the presence of chlorophyll *a* in aquatic environments. *Limnol. Oceanogr.* 21, 462-467.

Mallette, M. F. (1969). Evaluation of growth by physical and chemical means. In *Methods in microbiology,* 1, (ed. J. R. Norris & D. W. Ribbons) 521-566. London. Academic Press.

Marker, A. F. H. (1972). The use of acetone and methanol in the estimation of chlorophyll in the presence of phaeophytin. *Freshwat. Biol.* 2, 361-385.

Mayfield, C. I. (1975). A simple fluorescence technique for *in situ* soil microorganisms. *Can. J. Microbiol.* 21, 727-729.

Melchiorri-Santolini, U. (1972). Enumeration of microbial concentration in dilution series (MPN). In *Techniques for the assessment of microbial production and decomposition in fresh waters* (ed. Y. I. Sorokin and H. Kadota) IBP Handbook No. 23, 64-70. Oxford. Blackwell Scientific Publications.

Meyer-Reil, L.-A. (1978). Autoradiography and epifluorescence microscopy combined for the determination of number and spectrum of actively metabolizing bacteria in natural waters. *Appl. environ. Microbiol.* 36, 506-512.

*Meynell, G. G. & Meynell, E. (1970). *Theory and practice in experimental bacteriology.* 2nd edn. Cambridge. University Press.

Michiels, M. (1974). Biomass determination of some freshwater ciliates. *Biol. Jaarb.* 42, 132-136.

Miles, A. A. & Misra, S. S. (1938). The estimation of the bactericidal power of the blood. *J. Hyg., Camb.* 38, 732-749.

Miller, C. A. & Vogelhut, P. O. (1978). Chemiluminescent detection of bacteria: experimental and theoretical limits. *Appl. environ. Microbiol.* 35, 813-816.

*Ministry of Housing & Local Government (1969). *The bacteriological examination of water supplies.* Reports on Public Health and Medical Subjects, No. 71. London. HMSO.

Moran, P. A. P. (1954). The dilution assay of viruses. *J. Hyg., Camb.* **52**, 444-462.

Moriarty, D. J. W. (1977). Improved method using muramic acid to estimate biomass of bacteria in sediments. *Oecologia*, **26**, 317-323.

Munro, A. L. S. & Brock, T. D. (1968). Distinction between bacterial and algal utilization of soluble substances in the sea. *J. gen. Microbiol.* **51**, 35-42.

Niemelä, S. (1965). The quantitative estimation of bacterial colonies on membrane filters. *Suomal. Tiedeakat. Toim.* A4, **90**, 1-63.

Nissenbaum, G. (1953). A combined method for the rapid fixation and adhesion of ciliates and flagellates. *Science, N.Y.* **118**, 31-32.

Norman, R. L. & Kempe, L. L. (1960). Electronic computer solution for the MPN equation used in the determination of bacterial populations. *J. biochem. microbiol. technol. Engng,* **2**, 157-165.

*Norris, J. R. & Ribbons, D. W. (Eds) (1969-1976). *Methods in microbiology,* 1-9. London. Academic Press.

Norris, K. P. & Powell, E. O. (1961). Improvements in determining total counts of bacteria. *Jl R. microsc. Soc.* **80**, 107-119.

Oleniacz, W. S., Pisano, M. A., Rosenfield, M. H. & Elgart, R. L. (1968). Chemiluminescent method for detecting micro-organisms in water. *Environ. Sci. Technol.* **2**, 1030-1033.

Olson, F. C. W. (1950). Quantitative estimates of filamentous algae. *Trans. Am. microsc. Soc.* **59**, 272-279.

Overbeck, J. (1974). Microbiology and biochemistry. *Mitt. int. Verein. theor. angew. Limnol.* **20**, 198-228.

Patriquin, D. G. & Dobereiner, J. (1978). Light microscopy observations of tetrazolium-reducing bacteria in the endorhizosphere of maize and other grasses in Brazil. *Can. J. Microbiol.* **24**, 734-742.

Paerl, H. W. (1974). Bacterial uptake of dissolved organic matter in relation to detrital aggregation in marine and freshwater systems. *Limnol. Oceanogr.* **19**, 966-972.

Paerl, H. W. (1977). Bacterial sediment formation in lakes: trophic implications. In *Interactions between sediments and fresh water* (ed. H. L. Golterman) 40-47. The Hague. W. Junk.

Paerl, H. W. & Goldman, C. R. (1972). Heterotrophic assays in the detection of water masses at Lake Tahoe, California. *Limnol. Oceanogr.* **17**, 145-148.

Paerl, H. W. & Lean, D. R. S. (1976). Visual observations on phosphorus movement between algae, bacteria and abiotic particles in lake waters. *J. Fish. Res. Bd Can.* **33**, 2805-2813.

Paerl, H. W. & Shimp, S. L. (1973). Preparation of filtered plankton and detritus for study with scanning electron microscopy. *Limnol. Oceanogr.* **18**, 802-805.

Paerl, H. W. & Williams, N. J. (1976). The relation between adenosine triphosphate and microbial biomass in diverse aquatic ecosystems. *Int. Revue ges. Hydrobiol. Hydrogr.* **61**, 659-664.

*Page, F. C. (1976). An illustrated key to freshwater and soil amoebae. *Scient. Publs. freshwat. biol. Ass.* No. 34, 155 pp.

Paton, A. M. & Jones, S. M. (1973). The observation of micro-organisms on surfaces by incident fluorescence microscopy. *J. appl. Bact.* **36**, 441-443.

Paton, A. M. & Jones, S. M. (1975). The observation and enumeration of microorganisms in fluids using membrane filtration and incident fluorescence microscopy. *J. appl. Bact.* **38**, 199-200.

Perfilev, B. V. & Gabe, D. R. (1969). *Capillary methods of investigating microorganisms.* Edinburgh. Oliver & Boyd.

Peroni, C. & Lavarello, O. (1975). Microbial activities as a function of water depth in the Ligurian Sea: an autoradiographic study. *Mar. Biol.* **30**, 37-50.

Pike, E. B. & Carrington, E. G. (1972). Recent developments in the study of bacteria in the activated-sludge process. *J. Inst. Wat. Pollut. Control,* **6**, 1-23.

Pike, E. B., Carrington, E. G. & Ashburner, P. A. (1972). An evaluation of procedures for enumerating bacteria in activated sludge. *J. appl. Bact.* **35**, 309-321.

Pomeroy, L. R. & Johannes, R. E. (1968). Occurrence and respiration of ultraplankton in the upper 500 metres of the ocean. *Deep-Sea Res.* **15**, 381-391.

Postgate, J. R. (1969). Viable counts and viability. In *Methods in microbiology,* **1** (ed. J. R. Norris & D. W. Ribbons), 611-628. London. Academic Press.

Pugsley, A. P. & Evison, L. M. (1975). A fluorescent antibody technique for the enumeration of faecal streptococci in water. *J. appl. Bact.* **38**, 63-65.

Ramsay, A. J. (1974). The use of autoradiography to determine the proportion of bacteria metabolising in an aquatic habitat. *J. gen. Microbiol.* **80**, 363-373.

Ramsay, A. J. & Fry, J. C. (1976). Response of epiphytic bacteria to the treatment of two aquatic macrophytes with the herbicide paraquat. *Wat. Res.* **10**, 453-459.

Razumov, A. S. (1947). *Methods of microbiological studies of water.* Moscow. Vodgeo.

Reynolds, C. S. & Jaworski, G. H. M. (1978). Enumeration of natural *Microcystis* populations. *Br. phycol. J.* **13**, 269-277.

Ride, J. P. & Drysdale, R. B. (1972). A rapid method for the chemical estimation of filamentous fungi in plant tissue. *Physiol. Plant Path.* **2**, 7-15.

Riemann, B. (1978). Absorption coefficients for chlorophylls a and b in methanol and a comment on interference of chlorophyll b in determinations of chlorophyll a. *Vatten,* **3**, 187-194.

Rogers, C. J., Chambers, C. W. & Clarke, N. A. (1967). Spectrofluorimetric determination of nanogram amounts of a, ε-diaminopimelic acid (2,6-diaminoheptanedoic acid) a bacterial cell wall constituent. *Analyt. Biochem.* **20**, 321-324.

*Rosswall, T. (Ed.) (1973). *Modern methods in the study of microbial ecology. Bull Ecol. Res. Comm. (Stockholm),* **17**. Swedish Natural Science Research Council.

Rotman, B. & Papermaster, B. W. (1966). Membrane properties of living mammalian cells as studied by enzymatic hydrolysis of fluorogenic esters. *Proc. Natn. Acad. Sci. U.S.A.* **55**, 134-141.

Rowe, R., Todd, R. & Waide, J. (1977). Microtechnique for most-probable-number analysis. *Appl. environ. Microbiol.* **33**, 675-680.

Saunders, G. W. (1972). The transformation of artificial detritus in lake water. *Memorie Ist. ital. Idrobiol.* **29** Suppl., 261-288.

Schmidt, E. L. (1973). Fluorescent antibody techniques for the study of microbial ecology. *Bull. Ecol. Res. Comm. (Stockholm)*, **17**, 67-76.

Setaro, F. & Morely, C. G. D. (1976). A modified fluorometric method for the determination of microgram quantities of DNA for cell or tissue cultures. *Analyt. Biochem.* **71**, 313-317.

Sharabi, N. El-D. & Pramer, D. (1973). A spectrophotofluorometric method for studying algae in soil. *Bull. Ecol. Res. Comm. (Stockholm)*, **17**, 77-84.

Sharma, P. D., Fisher, P. J. & Webster, J. (1977). Critique of the chitin assay technique for estimation of fungal biomass. *Trans. Br. mycol. Soc.* **69**, 479-483.

Sheldon, R. W. (1972). Size separation of marine seston by membrane and glass-fiber filters. *Limnol. Oceanogr.* **17**, 494-498.

*Siegel, S. (1956). *Non-parametric statistics for the behavioral sciences.* New York. McGraw-Hill.

*Skerman, V. B. D. (Ed.) (1969). *Abstracts of microbiological methods.* New York. Wiley Interscience.

Slovacek, R. E. & Hannan, P. J. (1977). In vivo fluorescence determinations of phytoplankton chlorophyll *a*. *Limnol. Oceanogr.* **22**, 919-925.

*Snedecor, G. W. & Cochran, W. G. (1967). *Statistical methods.* 6th edn. Iowa. State University Press.

*Sokal, R. R. & Rohlf, F. J. (1969). *Biometry.* San Francisco. W. H. Freeman & Co.

*Sorokin, Y. I. & Kadota, H. (Eds) (1972). *Techniques for the assessment of microbial production and decomposition of fresh waters.* IBP Handbook No. 23. Oxford. Blackwell Scientific Publications.

*Stainton, M. P., Capel, M. J. & Armstrong, F. A. J. (1977). The chemical analysis of fresh water. 2nd ed. *Misc. Spec. Publs Fish. mar. Serv. Can.* **25**, 166 pp.

Stanier, R. Y. & Smith, J. H. C. (1960). The chlorophylls of green bacteria. *Biochim. biophys. Acta*, **41**, 478-484.

Stanley, P. M. & Staley, J. T. (1977). Acetate uptake by aquatic bacterial communities measured by autoradiography and filterable radioactivity. *Limnol. Oceanogr.* **22**, 26-37.

Stein, J. R. (Ed.) (1973). *Handbook of phycological methods. Culture methods and growth measurements.* Cambridge. University Press.

Stevenson, L. H. & Colwell, R. R. (Eds) (1973). *Estuarine microbial ecology.* Columbia. University of South Carolina Press.

Strayer, R. F. & Tiedje, J. M. (1978). Application of fluorescent-antibody technique to the study of a methanogenic bacterium in lake sediments. *Appl. environ. Microbiol.* **35**, 192-198.

*Strickland, J. D. H. & Parsons, T. R. (1968). A practical handbook of seawater analysis. *Bull. Fish. Res. Bd Can.* 167, 310 pp.

Strugger, S. (1948). Fluorescence microscope examination of bacteria in soil. *Can. J. Res.* Ser. C. 26, 188-193.

Swift, M. J. (1973a). The estimation of mycelial biomass by determination of the hexosamine content of wood tissue decayed by fungi. *Soil Biol. Biochem.* 5, 321-322.

Swift, M. J. (1973b). Estimation of mycelial growth during decomposition of plant litter. *Bull. Ecol. Res. Comm. (Stockholm)*, 17, 323-328.

Takahashi, M. & Ichimura, S. (1968). Vertical distribution and organic matter production of photosynthetic sulfur bacteria in Japanese lakes. *Limnol. Oceanogr.* 13, 644-655.

Talling, J. F. & Driver, D. (1963). Some problems in the estimation of chlorophyll *a* in phytoplankton. *Proc. Conference of Primary Productivity Measurement, Marine and Freshwater*, Hawaii, 1961. U.S. Atomic Energy Comm. TID-7633, 142-146.

Taylor, C. B. (1940). Bacteriology of freshwater. I. Distribution of bacteria in English Lakes. *J. Hyg., Camb.* 40, 616-640.

Taylor, J. (1962). The estimation of numbers of bacteria by tenfold dilution series. *J. appl. Bact.* 25, 54-76.

Tracey, M. V. (1955). Chitin. In *Modern methods of plant analysis*, 2, (ed. K. Paech & M. V. Tracey), 264-274. Berlin, Springer Verlag.

Trolldenier, G. (1973). The use of fluorescence microscopy for counting soil micro-organisms. *Bull. Ecol. Res. Comm. (Stockholm)*, 17, 53-59.

Trüper, H. G. & Yentsch, C. S. (1967). Use of glass fiber filters for the rapid preparation of *in vivo* absorption spectra of photosynthetic bacteria. *J. Bact.* 94, 1255-1256.

*Udenfriend, S. (1962). *Fluorescence assay in biology and medicine.* New York. Academic Press.

*Udenfriend, S. (1969). *Fluorescence assay in biology and medicine.* Vol. 2. New York. Academic Press.

Utermöhl, H. (1958). On the perfecting of quantitative phytoplankton methods. *Mitt. int. Verein theor. angew. Limnol.* 9, 1-38.

*Vollenweider, R. A. (Ed.) (1974). *A manual on methods for measuring primary production in aquatic environments.* IBP Handbook No. 12. 2nd edn. Oxford. Blackwell Scientific Publications.

Waid, J. S., Preston, K. J. & Harris, P. J. (1973). Autoradiographic techniques to detect active microbial cells in natural habitats. *Bull. Ecol. Res. Comm. (Stockholm)*, 17, 317-322.

Watson, S. W., Novitsky, T. J., Quinby, H. L. & Valois, F. W. (1977). Determination of bacterial number and biomass in the marine environment. *Appl. environ. Microbiol.* 33, 940-946.

Weaver, R. W. & Zibilske, L. (1975). Affinity of cellular constituents of two bacteria for fluorescent brighteners. *Appl. Microbiol.* 29, 287-292.

Wetzel, R. G. & Otsuki, A. (1974). Allochthonous organic carbon of a marl lake. *Arch. Hydrobiol.* 73, 31-56.

Wetzel, R. G. & Westlake, D. F. (1974). Periphyton. In *A manual on methods for measuring primary production in aquatic environments*. IBP Handbook No. 12. 2nd edn (ed. R. A. Vollenweider), 42-44. Oxford. Blackwell Scientific Publications.

Wiebe, W. J. & Bancroft, K. (1975). Use of adenylate energy charge ratio to measure growth state of natural microbial communities. *Proc. natn. Acad. Sci. U.S.A.* **72,** 2112-2115.

Willoughby, L. G. (1978). Methods for studying micro-organisms on decaying leaves and wood in fresh water. In *Techniques for the study of mixed populations* (ed. D. W. Lovelock & R. Davies). *Soc. appl. Bact. tech. Ser.* No. 11, 31-50. London. Academic Press.

Wilson, A. L. (1973). The performance characteristics of analytical methods. III. *Talanta,* **20,** 725-732.

*****Winberg, G. G. & collaborators (Eds) (1971).** *Symbols, units and conversion factors in studies of fresh water productivity.* International Biological Programme. London. IBP Central Office.

Woelkerling, W. J., Kowal, R. R. & Gough, S. B. (1976). Sedgewick-Rafter cell counts: a procedural analysis. *Hydrobiologia,* **48,** 95-107.

Young, M. R. & Smith, A. U. (1964). The use of euchrysine in staining cells and tissues for fluorescence microscopy. *Jl R. microsc. Soc.* **82,** 233-244.

Zimmermann, R., Iturriaga, R. & Becker-Birck, J. (1978). Simultaneous determination of the total number of aquatic bacteria and the number thereof involved in respiration. *Appl. environ. Microbiol.* **36,** 926-935.

INDEX

Acid digestion, 12, 14, 17
Acridine orange, 28, 48
Actinomycetes,
  cell walls, 83
  colony counts, 54
  sample handling, 15, 16
Adenosine triphosphate, see ATP
Adenylate energy charge, 71, 72
Amann's lactophenol, 18
Amoebae, colony counts, 55
Anaerobic conditions, 11, 16
Anoxic water
  ATP in, 66
  filtration, 11
  quenching of fluorescence by, 74, 87
ATP, 14
  analysis and biomass estimates, 70, 71
  enzyme preparation, 67, 68
  extraction, 65-67
  light emission, 68-70
Autoradiography, 45
Autotrophic carbon fixation, 45

Bacteriochlorophyll
  fluorimetric analysis, 81
  interference in chlorophyll a
    determination, 78
  spectrophotometric analysis, 80
Bacteriophage, 56
Bdellovibrio, 56
Beggiatoa, 16
Benthos,
  definition of, 9
  fungal chitin in, 82
  muramic acid in, 85
  sample handling, 15, 16
  summary guide to direct counts in, 49, 50
  use of counting chamber, 33 *et seq.*
Biochemical analysis,
  analysis of data, 88
  sample filtration for, 11
Bright field illumination, see Microscopy

Carbon analysis, 64
Casitone-glycerol-yeast extract agar, 54
Cellulose decomposers, 51
Cellulosic membranes, see Membrane
  filters
Chemical analysis
  analysis of data, 88
Chemiluminescence, luminol, 87
Chemolithotrophs, 52
Chitin determination, 81, 83
Chitinase, 83
Chlorobiaceae, fluorescence analysis, 81
Chlorophyll a, 76
  fluorescence microscopy, 29
  fluorimetric analysis, 78-80
  spectrophotometric analysis, 77-78
Ciliates,
  analysis of counts, 44, 45
  direct counts, 39, 40, 42
  permanent preparations, 18
Cocconeis, 15
Colloidal chitin agar, 54
Contagious distribution, 92, 93
Culture media, 54, 55
Cyanobacteria
  fluorescence, 26
  muramic acids in, 84

Density separation by colloidal silica, 15
Deoxyribonucleic acid, see DNA
Dilution
  and distribution of viable counts, 56
  and MPN methods, 57
  for plate counts, 52
  of samples, 11 *et seq.*
Diurnal migration, 14
DNA, analysis for, 72 *et seq.*

EDTA, use in cell lysis, 76
Ehrlich's reagent, 82, 83
Electron microscopy
  permanent preparations for, 17
  use of polycarbonate membranes for, 20

Endotoxin, *see* Lipopolysaccharide
Epifluorescence microscopy (*see also* Microscopy)
  black membrane filters for, 20, 21
  fluorescent antibody technique, 47, 48
  microscopical techniques, 25-29
  with microautoradiography, 47
Erythrosine staining
  of bacteria on membranes, 25
  of bacteria in counting chambers, 43
  of bacterial colonies, 53
Esterase and fluorescence microscopy, 26
Eukaryotic algae, fluorescence microscopy, 26
Exospores, and sample pretreatment, 15

Faecal bacteria
  culture media and methods for, 55
  membrane filter counts, 53
Ferric iron
  and filtration, 67
  precipitation, 11
Filamentous fungi, biomass estimates, 74
Filters, (*see also* Membrane filters)
  copper, nickel, nylon, stainless steel, 14
Firefly luciferase, *see* ATP
Fluorescein diacetate, viability test, 26, 49
Fluorescein isothiocyanate, 47
Fluorescence, *see* Microscopy *and* Spectrofluorimetric analysis
Fluorimetry, fluorimetric analysis, *see* Spectrofluorimetric analysis
Fluorochromes, 27, (*see also* Microscopy)
Formazans in cytochemical analysis, 48
Freeze-thawing and cell lysis, 76
Fuchs-Rosenthal haemocytometer, 33
Fungal spores, as internal standards, 42
Fungi Imperfecti, growth media for, 55

Glucosamine,
  in the measurement of fungal biomass, 81, 82
Glutaraldehyde
  as an inhibitor, 46
  for clearing cellulosic membranes, 24
  permanent preparations for electron microscopy, 17

Haptobenthos,
  definition of, 9
  sample pretreatment, 15, 16

Heine condenser, 25 (*see also* Microscopy)
Helber chamber,
  for direct counts, 33, 34
  microscopy, 43
Herpobenthos,
  definition of, 9
  sample pretreatment, 14
Heterotrophs, viable counts of, 53 *et seq.*
Homogenization, sample pretreatment by, 15
Hydrogen sulphide in chlorophyll extraction, 77
Hyphae,
  biomass estimates of, 74
  the effect of ultrasonication on, 10
Hyphomycetes, cell wall constituents of, 81

Incident light microscopy, *see* Microscopy
Infra-red analysis of carbon, 64
Irgalan black for dyeing membrane filters, 21
Iron bacteria, examination on membrane filters, 25
Isotopes and autoradiography, 45 *et seq.*

Limulus polyphemus, see Lipopolysaccharide assay
Lipopolysaccharide assay, estimation of bacterial numbers from, 86
Logarithmic distribution, 44
Log-normal distribution, 56
Luciferase, *see* ATP
Luciferin, *see* ATP
Lugol's solution,
  permanent preparations of algae, 17
  sedimentation of algae, 38
  use in autoradiography, 46
Lycoperdon as a source of chitinase, 83
Lysine in bacterial cell walls, 83
Lysozyme, and cell lysis, 76

Magnetic stirrer,
  and sub-sampling, 36
  in sample concentration, 12
Meiobenthos,
  density separation, 15
  direct counts, 40
Membrane filters,
  cellulosic, 11, 14, 19-21, 24, 64, 66, 67, 80
  polycarbonate, 11, 12, 19, 20-22, 24, 64, 87

INDEX

Mercury burners in fluorescence microscopy, 26
Methyl fluorescein phosphate and fluorescence microscopy, 28
Methylene blue, use in
  colony counts of bacteria on membranes, 53
  direct counts of bacteria on membranes, 25
Microautoradiography 45, 46, 47
Microcystis, counts of individual cells, 14
Microinterferometer, use in chamber counts, 36
Microscopy,
  bright field, 24, 25, 33, 43, 49, 53
  fluorescence, 17, 21, 25-29, 40, 43, 47, 48
  incident light, 15, 19, 38, 47
  phase contrast, 25, 34, 36, 42, 43, 49, 53, 54
  scanning electron, 20, 21
  stereo, 53
  transmission electron, 21
  transmitted light 15, 19, 20, 24, 26, 38
Most Probable Number counts,
  analysis of results, 59
  media for, 54, 55
  procedure, 57, 58
Mucopeptides in bacteria, 83
Muramic acid in prokaryotes, 83, 86
Mycelium,
  biomass estimates of, 74
  glucosamine content of, 82
  sample pretreatment, 15, 16

Negative binomial distribution, 44
Neuston, 9
Newton's rings and chamber counts, 34
Nitrogen analysis, 64
Non-parametric statistical tests, 9, 44, 94
Normal distribution, 92, 94
Null hypothesis, 90, 94

Obligate anaerobes, ATP analysis of, 67
Oligotrophic waters,
  ATP analysis of, 66
  concentration of samples from, 11
Organic matter,
  digestion of, 12
  interference in fluorescence microscopy, 48

Parametric statistical tests, 30, 44, 88, 92, 94
Penicillin, use in viable counts of fungi, 55
Periphyton, 9
Petroff-Hauser chamber, 33, 34
Phase-contrast microscopy, *see* Microscopy
Phosphatase,
  and fluorescence microscopy, 28
  viable counts of bacteria containing, 54
Phosphorus, inorganic and membrane filtration, 11
Phosphorus pentoxide as a desiccant, 64
Phycomycetes,
  aquatic, viable counts, 55
  parasitic, concentration for direct counts, 16
Plankton,
  definition of, 9
  sample handling, 15, 16
  summary guide to direct counts of, 49, 50
  use of counting chamber for, 33 *et seq.*
Plate counts,
  analysis of results, 56, 57
  pretreatment for, 15
  procedure 52, 54, 55
Poisson distribution, 15, 30, 44, 56, 59, 60, 61
Polycarbonate membranes *see* Membrane filters
Production of heterotrophs, 47

Quartz-iodine for fluorescence microscopy, 26

Random distribution, 31, 92, 93 (*see* Poisson distribution)
Regular distribution, 92, 93
Rhodospirillaceae, fluorescence analysis, 81
Rotifers,
  chamber counts, 36
  direct counts on membranes, 25, 30

Scanning electron microscope, *see* Microscopy
Sedgewick-Rafter cell, 39, 44
Sedimentation chambers,
  analysis of results from, 44
  for counting algae, 36, 38
Seston, analysis of, 63-65
Sonication, *see* Ultrasonic disruption

Spectrofluorimetric analysis
  of bacteriochlorophylls, 81
  of chlorophyll *a,* 78-80
  of DNA, 73
Spores,
  and sample pretreatment, 15, 16
  and viable counts, 55
Standing crop, 8
Stereo microscopy, *see* Microscopy
Sub-sampling, analysis of, 91
Sudan black B for dyeing membrane filters, 21
Sonication, *see* Ultrasonic disruption

Tetrazolium salts,
  in cytochemistry, 48, 49
  in viable counts of bacteria, 55
Thin-layer chromatography,
  in chlorophyll analysis, 78
  in muramic acid analysis, 85
Transmission electron microscopy, *see* Microscopy

Transmitted light microscopy, *see* Microscopy
Tritium in microautoradiography, 45
Trolldenier slide, 40, 43

Ultrasonic distruption in
  extraction of bacteriochlorophyll, 80
  extraction of chlorophyll *a,* 76, 78
  sample pretreatment, 14-16

Vacuum filtration, 22
Viable counts
  pretreatment of samples, 16
  procedure, 51 *et seq.*

Xenon burners in fluorescence microscopy, 26

Yeasts, direct counts with fluorescein diacetate, 26

Zernicke condenser, 25
  see also Microscopy.